Seed Pathology

Dr. Gurnam Singh, a Ph.D. from Narendra Deva University of Agriculture and Technology, Faizabad, has worked with Seed Certification wing of UP State Agriculture Department in various capacities. He is a fellow of the two premier science academics of India and member of several professional societies. He has published 17 research and review papers in internationally reputed journals, besides five books on various aspects of seed science. He visited many countries, attended seminars, workshops and corporate conferences of National and International status.

Seed Pathology

Editor

Gurnam Singh

2025

Gene-Tech Books

ISBN10 81-89729-92-6
ISBN13 978-81-89729-92-9

Published by : **GENE-TECH BOOKS**
4762-63/23, Ansari Road, Darya Ganj,
NEW DELHI - 110 002
Phone: 43003222
e-mail: genetechbooks@yahoo.co.in

PRINTED IN INDIA

Preface

Ever since Paul Neergard - the widely regarded 'Father of Seed Pathology' introduced the subject as a separate and important agricultural branch, it has made immense strides in the field of plant pathology. A relatively new biological subfield, this science looks all set to usher in a new era of development in botany.

This science makes a comprehensive study of the seed borne diseases and pathogens. Mechanisms of infection, seed transmission, role of seed-borne inocula in seed development, techniques for detection of seed-borne pathogens and non-pathogens, seed certification standards - all these make up for the curriculum of the subject. Besides, diseases caused by bacteria, fungi, viruses, nematodes, viroids, etc., and the physiological and mechanical disorders of seeds, are also important factors contributing to the study.

This book seeks to facilitate the reader to grasp the basic principles of seed pathology. With the immense potential of seed science and technology to made huge contributions to the future of agricultural sector, seed pathology should be all set to gain from such a scientific revolution. Bringing forth a comprehensive analysis of the subject, it is hoped that this manual fosters an enriching learning experience for the reader.

Gurnam Singh

Contents

1

Seed Pathology

In the past, damping-off fungi, cone rust, a number of cone and seed insects in pines, and fungal damages to oak acorns, hazelnuts, chestnuts, walnuts, and seeds of birch and elm were considered the only major problems in seed and seedling production. Many fungi have now been isolated and studied for their effects on seeds of both conifer and hardwood tree species and some have caused considerable losses of seeds and seedlings.

According to Chalermpongse and others, in Thailand, the loss of seeds due to infection of *Botryodiplodia theobromae* on *Swietenia macrophylla* was 92 percent, *Colletotrichum gloeosporoides* on *Dalbergia cochinensis* was 4 percent, *Alternaria longissima* on *Bauhinia* sp. was 2 percent, *Pestalotiopsis* sp. on *Cassia bakeriana* was 6 percent, *Macrophoma* sp. on *Eucalyptus camaldulensis* was 1 percent, and *Fusarium* sp. on *Shorea obtusa* was 2 percent.

Deterioration of tree seeds by fungi involves problems differing in many aspects from those of grains. For example, tree seeds are exposed to many conditions before storage that permit the development of mold fungi. Cones are often collected in sacks and kept at collection points or processing plants for varying periods under conditions that favour fungal development prior to seed extraction. After extraction, the seeds are dried and stored for varied periods until used in field or nursery. Hence, it is important to know the characteristics of fungi associated with important species, what damage they cause, where and when and under what circumstances the damage occurs, and what can be done to prevent the damage.

The science of tree seed pathology is still very young. The occurrence and distribution of most of the tree seed pathogens is well realised but there is very little understanding of their impact on seed production, seed quality, and seed viability. Interest in these problems has been growing steadily in the past few decades, but recently this interest has changed into serious concern mostly because of the problems encountered in the renewal and management of the forests, which are a major resource worldwide. Some information about the characteristics of seed-borne fungi is now available but is mostly related to temperate tree species.

Tree Seeds Fungi

Fungi associated with tree seeds vary in different host species, in different regions, and in different years. Many of them are molds and develop on the seed surface only; some cause internal infections too. Nearly all seeds carry spores of various microscopic fungi either on the surface or within the seed. A superficial mycoflora is almost always found because of the ready adhesion of spores to the uneven surface of the seeds. Although the number of spores occurring varies considerably, as in Norway spruce and Scotch pine, it can be as high as 50 to 150,000 spores, and, in some seedlots, several hundred thousand spores per 1 g seed.

Under favourable conditions, some spores germinate, the mycelium penetrating into the cotyledons of the seed and feeding on the embryo. Several kinds of fungi can be associated with tree seeds. Thus, there are species causing decay and reducing the germination of stored seeds, species attacking germinating seeds and seedlings, and other species that are more or less harmless, or at least appear to be so. Present knowledge does not permit precise separation of individual species of fungi occurring on seeds.

However, it is clear that many species usually considered as unimportant and harmless can cause considerable damage under certain conditions, for example: unsuitable storage conditions, seeds of poor quality (immature, low vigor, or heavily molded), unsuitable growing conditions (involving moisture, temperature, or aeration), etc. Therefore, while evaluating the importance of these fungi, it is essential to consider the biology of the individual species of fungi. Depending upon their location, the seed-borne fungi can, in general, be classified in two groups: externally seedborne and internally seed-borne.

The first group includes species of *Botryosphaeria, Botrytis, Fusarium, Mucor, Phialophora, Rhizopus*, and *Trichothecium*. They are not usually host specific and may involve more than one species. Some of the wellknown internally seed-borne fungi include species of *Alternaria, Aspergillus, Botrytis, Botryodiplodia, Caloscypha, Cephalosporium, Fusarium, Phoma, Schizophyllum, and Sirococcus*. These may cause deterioration of seed quality and pre- or post emergence mortality of seedlings.

Urosevic provided instructions for health testing of oak acorns, including a key for distinguishing fungi from acorns, and divided the acorn mycoflora into two groups:

(i) parasites and semiparasites such as *Ciboria batschiana*, *Ophiostoma* spp., *Gloeosporium quercinum*, *Phomopsis quercella*, *Cytospora intermedia*, *Botrytis cinerea*, and *Pestalotia* sp.; and

(ii) sparophytes including *Alternaria*, *Aspergillus*, *Fusarium*, *Penicillium*, *Trichoderma*, and others.

Based upon their pathogenicity, Sutherland classified the seed-borne fungi of conifers as

(i) saprophytes or weak pathogens;

(ii) pathogens such as the cold fungus *Caloscypha fulgens* which consistently kills seeds;

(iii) pathogens mainly important as seedling pathogens, e.g. *Sirococcus conigenus*; and

(iv) fungi, e.g. *Fusarium* spp., whose pathogenicity depends upon factors including fungus species and pathogenic strain, and host and host stress.

Seed-Borne Fungi

Symptoms of seed-borne diseases are usually divided into pre and post emergence damping-off. The former consists of reduced emergence and decay of the radicle just emerged from the seedcoat; the latter is subdivided into root rot, cotyledon rot, and basal stem rot after the seedlings emerge from the soil. Reduction in seed germination, decay and loss of viability of seeds during storage, and the diseases of seedlings are among the major problems wrought by fungal pathogens.

Reduction of Seed Germination

Inhibition of coniferous seed germination by widespread contaminators and through artificially inoculated fungi has been reported. Huss observed that the molding had virtually no effect on pine seeds of high viability, but poor quality seeds suffered a substantial reduction of germination. It has also been observed that the destruction of seeds largely depended on their growth rate, and as germination progressed, the resistance to destruction increased. The extremely common and numerous species of mold fungi, viz. species of *Mucor, Rhizopus, Trichothecium, Botrytis, Penicillium,* and others, which colonised the surface of *Quercus acorns* or got into the surface tissues, were found of secondary importance in the loss of germinability of acorns.

Gibson, however, reported that the saprophytic fungi, viz. *Aspergillus* spp., *Mucor* spp., *Rhizopus* sp., *Trichoderma* sp., and *Cladosporium* sp., of the seedcoat microflora could, under favourable conditions, invade tissues of the germinating seeds and kill the seedlings of *Pinus patula*. The seedcoat microflora could thus be directly responsible for the weakening of seed vigor, predisposing it to the attack of soil-borne pathogenic fungi. Shea supported the view and mentioned that the influence of molds on seeds could vary considerably and their mere presence did not mean that they were harmful.

However, Prisyazhnyuk mentioned that the greater the infection by fungi of seeds, the lower the germinability of seeds of *Pinus sylvestris, Larix sibirica, Picea abies,* and *Abies sibirica*. Rowan and De Barr observed extensive molding of three seed lots of slash pine during germination tests. Following the standard testing procedures, the seeds were cracked individually and almost 90 percent of them were found full, although the germination ranged from 31 to 79 percent only. Fusarium solani was obtained from the ungerminated seeds.

Out of the 12 *Fusarium* isolates tested through inoculation on Pinus patula seeds, 3 reduced germination, whereas 9 increased percent damping-off but did not affect seed germination or seedling growth. Leucaena seeds infected by *Colletotrichum graminicola* failed to germinate, and if the infection was carried to the nursery, seedlings under moisture stress succumbed to the damping-off disease. *Botryodiplodia theobromae*, which causes black dry rot in mahagony (Swietania macrophylla) and caused 92 percent seed deterioration in Thailand, did not affect germination but, after potting in the nursery, the rot developed during the hardening-off period of seedlings.

Stratification, also known as moist cold prechilling of seeds, is commonly used to break dormancy in seeds, and to attain vigorous, speedy, maximum, and uniform germination for laboratory testing and green house and nursery sowing. It is a common practice for most conifer and several hardwood seeds. Sutherland reported spread of *Caloscypha fulgens* on seeds of several conifer species at low (3 to 5°C) temperatures of stratification. The fungus mummifies the seeds, resulting in poor germination. Mittal and others observed the development and spread of fungi on *Pinus strobus* seeds during stratification. These fungi, generally, did not lower seed germination but diseased the germinants.

Decay and Loss of Viability

Large quantities of high-quality seeds are required annually for artifical regeneration. In view of the lack of uniformity and predictability of cone and seed crops, bulk quantities of seeds are collected in good seed years and stored for use in intervening years to ensure a continuous supply of seeds for annual production of planting stock and for direct seeding. Storability of seeds is dependent upon temperature, time, relative humidity, and method of storage, as well as the moisture content of, and initial fungal inoculum on, seeds to be stored. Improper storage of cones, as reported by Shea, caused heating of cones as a result of biological activity and these cones suffered more damage by fungi.

Sixty percent of the Pinus sylvestris seeds stored in sacks; about 30 percent of seeds stored in boxes, bins, and tin drums; and only 10 percent stored in hermetically sealed vessels were infected with various fungi. *Quercus* acorns lost up to 70 percent of their germination capacity during storage due to fungal infections. Achenes of *Platanus occidentalis* stored at 2°C showed no loss in germinability even after 7 months at 20 and 30°C; however, germinability decreased and most fungi on achenes increased with increasing temperature, relative humidity, and time of storage.

In conifers, Lavender found no loss in germination capacity of *Pseudotsuga menziesii* seeds stored in cones up to 4 months at normal fall temperatures in an unheated warehouse. Bloomberg reported that seeds in *Pseudotsuga menziesii* cones stored for 225 days under operational conditions were free of diseases but, during germination tests after extraction, up to 56 percent of them became diseased. Rediske and Shea observed significant reduction in the same seed viability when freshly picked cones of high moisture content (60 percent) were sacked and stored in outside bins in the fall. However, full viability of these seeds could be retained for 3 years when stored in sealed cones at 0°F.

When the same seeds were stored in canvas bags, germination was somewhat reduced within 6 months, and severely reduced after 12 months of storage. Gordon advised refrigeration of extracted seeds of Pseudotsuga menziesii immediately after their removal from the cone, to inhibit further fungal activity within the seed. However, Rediske and Shea supported Schubert's observation on Pinus monticola seeds: that fungi remained active in pine seed at temperatures below the freezing point. There have been suggestions for storing tree seeds at subfreezing temperatures to maintain their germinability.

Cryopreservation of seed germplasm at or near the temperature of liquid nitrogen has the potential for reducing deterioration of seed to such a low level that essentially perpetual preservation can be achieved. However, based on artificial inoculations of *Fusarium sporotrichioides* and *Mucor hiemalis* on water-soaked *Pinus strobus* seeds before storing at -18, -80, -145, and -197°C for 35 days, Mittal and Wang inferred that storing tree seeds at ultra-low temperatures will not eliminate contaminating fungal pathogens.

Seedling Diseases

Damping-Off

Pre- and post emergence damping-off caused by various fungi are the most dangerous diseases affecting conifers as well as hardwood species. Quiniones reported establishment of *Fusarium solani,* a soil inhabitant, in the seeds of *Leucaena* and *Agathis* which caused post emergence damping off in the nursery and in the outplanted seedlings.

Blight diseases

Sirococcus blight caused by a seed-borne fungus *Sirococcus strobilinus* is an important disease of seedlings of several spruce and pine species, and of *Pseudotsuga menziesii* throughout the northern Temperate Zone (Sutherland 1985). In this case, the pathogen attacked very young seedlings, killing the primary needles from the base upward. Dead seedlings remained upright, and small, black pycnidia usually formed at the base of infected needles. Diseased seedlings usually occurred randomly, characteristic of seed-borne diseases.

Wilting

Another important disease transferred by seeds is the tracheomycosis wilting of plants. This symptom can be elucidated as a reaction of the host to the irritation by the parasite wherein the typical blocking of trachae by thalli, and a yellowish brown, rubber-like substance filling the adjacent parenchymatous cells, are produced.

Others

Reduced seedling height and leaf symptoms (chlorotic and necrotic lesions and malformed leaves) are also sometimes observed in seedlings raised from the fungus-inoculated seeds of *Acer saccharum, Picea glauca, and Pinus strobus.*

Seed-Borne Fungi Testing

Seed-borne fungi testing includes isolation and study of fungi during cone collection and processing, seed extraction, processing, storage, germination, and seedling growth. Unlike with agricultural crops, methods for health testing of most forest tree seeds have not been standardised, and the testing is usually done using normal procedures of moist blotter and agar plates.

Singh and Mathur have elaborated on the seed-health testing methods including direct observation, washing test and incubation methods (blotter and agar plate) for seeds, and seedling symptom test and growing-on test for seedlings. Some special methods like dilution plate method, ultrasound technique, isozyme patterns, seed tissue excision, seed sectioning, radiography, and ELISA technique have also been discussed. The ISTA recommendations for germination testing, which are more clearly available for agricultural crop seeds, are usually followed as standard practice. Seed size and, sometimes, unavailability of tree seeds in large quantities, makes it difficult to use large numbers of seeds in testing. Therefore, it is important to find out how many seeds of a tree species should be tested in how many replicates.

For many years it has been difficult to secure accurate, maximum germination of all viable seeds or achieve the true planting value of forest tree seeds. For example, the dormant nature of the seeds of Abies balsamea, A.fraseri, and other Abies species, which required a moist prechill treatment of 21 to 28 days or more at 3 to 5°C, together with fungal contamination and growth during the 2-month overall test duration, have been responsible for sometimes erroneous, erratic, or negative germination results.

Another problem generally encountered in seed testing involves seed pretreatment. For this purpose, Wall and several others used 0.1-percent mercuric chloride solution for 2 minutes, following the ISTA recommendations for seed-health testing of agricultural crop seeds. Wall also used 0.5-percent sodium hypochorite solution for 2 to 3 minutes for surface sterilisation of diseased red pine seedlings. Mittal recommended treatment with 2-percent sodium hypochlorite solution for 10 minutes for Picea glauca and Pinus strobus seeds for greenhouse sowing as well as for laboratory testing. Thus, there existed different opinions on the type and duration of treatment that seeds should receive before testing.

For testing pathogenicity through artificial inoculation of seeds with some seed-borne fungi, different methods for seed inoculations are employed. Several workers have attempted rolling of seeds on fresh fungal cultures, whereas others used spore suspensions. At several platforms, a controversy existed over the method of inoculation and the testing environment, which need to be standardised for different types of seeds. The reduction of germination under conditions of artificial infection does not correspond exactly to the reduction of germination that occurs under conditions of natural infection.

Under natural conditions, the various microorganisms on seeds interact within themselves and with the microorganisms present in the soil or growing media. Such an interaction is often even antagonistic, which affects the ability of individual microorganisms to develop rapidly and to infect seeds. Often, in artificial inoculation studies, the conditions for facilitating microbial growth are provided. This suggests a need for testing the pathogenicity of various fungi in natural soils or growing media under greenhouse or field conditions.

Factors Affecting Spread of Seed-Borne Fungi

Abiotic Factors

Collection, extraction, and processing

The time, place, and method of collection of cones or seeds, and their subsequent handling during transport, extraction, and processing, affect the development and spread of mycoflora on tree seeds. It is generally presumed that cones and seeds acquire various fungi, including molds, while they are still on trees. Pinus pinea cones contained discoloured, powdery seeds, in some of which the kernel was still sound but in others was blackened and completely destroyed by the grayish mycelial growth of *Alternaria alternata*.

Similarly, a pathogen *Coniothyrium* sp., established itself in and on the Betula aleghanensis seed before it fell to the ground. However, Salisbury and Prisyazhnyuk reported that most of the individual seeds from tightly closed cones of conifers were completely free from molds. Seeds should be extracted from the cones immediately after harvest to minimise seed infection from the microbial population already present on cones. The later the oak acorns were collected from the fields, the more infected they were with fungi. *Penicillium, Fusarium, Alternaria,* and *Trichothecium roseum* were found to be the most common fungi during acorn development.

Development of *Cladosporium herbarum,* which was frequently encountered on the acorns in the first and last samples during their development, depended on the amount of rainfall; the highest occurrence was during the period of heavy rainfall. Dewinging of the seeds can lead to considerable reduction in germination, presumably through damage to the testa and subsequent fungal invasion. While studying the fungi associated with seeds of *Picea glauca* and *Pinus strobus* during cone processing and seed extraction, Mittal and Wang observed that the contamination occurred and spread during air drying and seed extraction processes, and that considerably more fungi occurred on both types of seeds after they were left on the forest floor for 15 days. Mojtahedi and others found that the wash water was a source of fungus contamination when fresh, uncracked pistachio nuts were tested for *Aspergillus flavus* and *aflatoxin* before and after a commercial washing treatment.

Storage conditions

The moisture content of seeds, initial fungal inoculum on

seeds, and the method of storage, all affect the spread of fungi on seeds during storage. Full viability of extracted Douglas fir seeds was retained for 3 years when stored in sealed cans at 0°F. However, when stored in canvas bags, germination was somewhat reduced within 6 months and seriously reduced after 12 months. Significant reduction in Douglas fir seed viability when freshly picked cones of high moisture content (60 percent) were sacked or stored in outside bins in the fall, was observed by Rediske and Shea. Immediate gentle drying or refrigeration maintained viability.

While orthodox seeds, which can tolerate low moisture content and low storage temperatures and, therefore, can be stored successfully for often longer periods, the recalcitrant tree seeds are fast-perishable. Their high moisture content and storage often at ambient or relatively higher temperatures help in establishing several storage fungi. Fast deterioration of the recalcitrant seeds by these storage fungi is expected to be due to the debilitation of seeds caused by internal moisture stress generated within the cells or tissues, primarily as a result of the water-requiring process of vacuolation. This debilitation impaired phytoalexin synthesis by the seeds, facilitating proliferation of associated fungi or bacteria.

Prevention of fungal activity during storage can be more easily achieved by controlling the moisture content of the seeds than by controlling the storage temperature, because fungal activity is possible between -8°C and +80°C when the seed moisture content and the relative humidity of storage are high enough. It is therefore important to find out the optimum moisture content and the storage temperature requirements for individual forest tree species.

Stratification

The killing of viable seeds of some conifer species has been particularly serious during the long, cold, prechill treatment at 3 to 5°C. Mittal and others studied the development and spread of fungi on Pinus strobus seeds during stratification, though the former did not, usually, lower the seed germination in further testing. The prechilled seeds germinated vigorously and speedily, and therefore probably escaped the damages, supporting the views of Gibson. However, some germinants failed to emerge completely from the seedcoat and others were damaged by top decays caused by *Alternaria alternata, Fusarium oxysporum,* and *Penicillium* variabile in the laboratory, possibly due to the high moisture content of the prechilled seeds and the high environmental temperature: the two important factors contributing to fungal development, spread, and infection. This suggested a need for treatment by some surface cleaning or sterilising agent before stratifying the seeds, especially the highly dormant seeds.

Cultural practices

Surface-sterilised seeds with pierced testa sowed in normal moist soil completely failed to germinate, presumably due to invasion by soil saprophytes. It seems possible that the destruction of seeds was facilitated by the proximity of a relatively large food base, the seed reserves adjoining the small volume of living tissue. Sitka spruce seeds incubated at 10°C were highly susceptible to fungal attack because they remained dormant while the pathogen was growing at near its maximum rate.

Losses in nurseries are not necessarily related to the time taken for seedlings to emerge, but they are likely to be greater in places where fluctuations in temperatures over 10°C are usually less prevalent. Losses are expected to increase with earlier dates of sowing. Autumn sowing is most unreliable because in warm soil, the seeds germinate early and escape damage, whereas in cold soil they do not germinate until the following spring, and suffer maximum loss.

Most soil mixes for containerised conifers contain vermiculite or perlite incorporated with sphagnum peat. This type of mix is usually well drained and acidic, the two factors that help reduce diseases. Major groups of pathogens associated with nursery diseases are species of Fusarium and watermolds, such as *Pythium* and *Phytophthora*. Although watermolds may be seed-borne, they are more often introduced into container nurseries through contaminated irrigation water. These fungi cause disease on very young seedlings, and are favoured by poorly drained soil mixes and prolonged wet conditions in the greenhouse.

Biotic Factors

Fungal colonisation of conifer seeds is facilitated by insects like seedbugs, and by squirrel damages to seeds. Elaborating on the insect-fungal interaction during infection of oak acorns, Urosevic reported that at the time of oviposition, various fungi were introduced by the insects into the acorns. These fungi then penetrated into the mature acorn tissues. Death of the acorns was thus brought about not only by the damage caused by the larvae of the weevil, but also by fungi accompanying the weevil. The penetration of the acorn by the fungal mycelium entailed negative effects more rapidly and more injurious than did the maturation feeding of the larvae. Acorns thus affected might represent a dangerous focus of infectivity during both initial storage and long-term storage of the acorns.

Control of Seed Pathogens

Seed Collection, Processing and Storage

Collection of seeds from healthy, disease-free areas or orchards; collection from healthy trees, healthy cones or acorns at the appropriate time; collection from tree and not from ground or from squirrel caches, etc.; transport of cones or seeds in wellaerated, clean, dry containers or bags; avoiding damage to seeds during extraction and processing; and use of optimum seed extraction and storage conditions, all need be studied for different forest tree species and considered for prevention of fungal infections on seeds.

Surface Treatment

Although adverse effects on seed germination have sometimes been reported, seed treatment with sterilants to reduce or eliminate fungal contamination has been considered necessary for production of healthy seedlings at several nurseries. For sterilising conifer seeds with minimal stimulation or retardation to them, an immersion of the seeds in a commercial detergent followed by treatment with 30-percent hydrogen peroxide has been recommended.

Hydrogen peroxide treatment (30 percent for 45 minutes) improved the total germination from 47 to 80 percent and from 25 to 61 percent in the 2, poor-quality, unstratified seeds of *Pinus taeda*. Water treatment at 57°C for 10 minutes was found quite effective in eliminating large numbers of seed-borne fungi of Pinus roxburghii and *P. wallichiana*. Delatour and others also suggested hot water (44°C for 8 hours) soak treatment for killing Ciboria batschiana in Quercus acorns.

Chemical Control

Coating seeds with a repellent against birds and small

rodents, and a fungicide against damping-off has been a common practice in forest tree nurseries at several places. Although a lot of literature on chemical seed treatment control of seed-fungi has accumulated, most studies were made on conifers The sulphuric acid treatment to Araucaria excelsa seeds, which has been prescribed by quarantine regulations against Cryptospora longispora, was found effective in eradicating the seed-borne fungi but the acid was detrimental to seed germination.

A 70 to 75 percent dust of PCNB applied to the seed of balsam, Fraser, and grand firs gave excellent control (100 percent) of Rhizoctonia solani without any injury to the germinating seedlings. Mittal and Sharma, based on their observations with different tree species (Cedrus deodara, Eucalyptus citriodora, E. hybrid, Pinus roxburghii, P. wallichiana, and Shorea robusta), suggested that Brassicol, Bavistin SD, and Dithane M-45, as seed dressers, could be used to effectively control most of the common seed-borne fungi of these tree species. For control of a common fungus, Aspergillus niger, on the seeds of Shorea robusta, seed treatment with Bavistin SD or Brassicol was most effective.

Effective control of several fungi, such as *Botryodiplodia theobromae, Colletotrichum gloeosporoides, Fusarium* spp., *Macrophomina phaseolina, Pestalotia sp., Phoma* sp., and *Phomopsis* sp., on the seeds of *Acacia auriculiformis, Albizzia* spp., *Gmellia arborea, Leucaena leucocephala,* several *Pinus* spp., *Pithecelobium dulce, Pterocarpus indicus, Cedrella odorata,* and *Grevillea robusta,* has been successfully achieved through the combined use of Benlate (0.15 percent) and Dithane M 45 (0.15 percent).

Seed treatments with oil, talc, and dye have also been found beneficial but much less so than treatment with Thiram (a fungicide) for Sitka spruce (Picea sitchensis).

Since detrimental effects of chemical seed treatment on seed germination and seedling quality have also been reported, it is desirable that lower concentrations, which should not be phytotoxic, be tried. Kozlowski reported that Captan at concentrations up to 2500 ppm did not affect seed germination of Pinus resinosa; however, concentrations of 500 ppm or higher injured roots, stems, and cotyledons within 13 days. Root injury consisted of collapse of root hair cells, epidermal cells, and cortical cells, and the cotyledon injury included the collapse of epidermal and mesophyll cells. Similar observations were made earlier by Cram and Vaartaja and Vaartaja.

Legislation

Vigorous implementation of the seed laws, such as the Seed Acts and Seed Certification Programmes for quality evaluation and management, and the standards for seed collection, extraction, storage, and movement, is needed to avoid the seed problems.

Practice

Australia is the origin of a unique and extensive resource of tree and shrub species which have proven to be of great value for the establishment of plantations in many parts of the world. Examples include approximately 5 million ha fastgrowing eucalypt plantations in Brazil; the Acacia mangium resource, almost 1 million ha, recently established in Indonesia; *Acacia saligna,* planted extensively in north Africa and the Middle East as a fodder tree; Acacia colei, planted around villages in semiarid Niger to provide edible seed to supplement inadequate diets, and the extensive plantings of Casuarina equisetifolia on sandy shorelines of southern China and Vietnam for typhoon protection and a wide range of timber and nonwood benefits.

Although Australian native trees, especially eucalypts, have been grown as exotics for more than a century, the area of plantations has expanded rapidly during the last 30 years. This expansion has been driven by the development of hardwood pulp as a major international commodity to meet the increasing demand for paper, and the widespread adoption of Australian trees for community forestry in Asia and parts of Africa. Having evolved in a continent characterised by climatic extremes and infertile soils, Australian native trees have proved to be well adapted to cultivation as multipurpose trees on degraded soils and provide a wide range of products including timber, poles, fuel, and oils.

The Australian Tree Seed Centre (ATSC), part of CSIRO Forestry and Forest Products, has acted for 35 years as a national seed bank, supplying seed to researchers in Australia and over 100 other countries. The seed originates in natural forests but over the last decade ATSC has been complementing these collections with seed-orchard seed. Seed orchards have been established in tropical and temperate Australia, and in several overseas countries in southern and Southeast Asia and Oceania, in collaboration with a wide range of agencies.

In 1987 research was initiated by ATSC to investigate the presence of fungal pathogens in stored seed. Although seed dispatched overseas by ATSC is routinely treated to meet the phytosanitary requirements of the recipient country, there was little information on the seed pathology of the three most important Australian native genera grown in plantations, domestically and overseas, namely Eucalyptus, Acacia, and Casuarina.

Storage Fungi and Seed-Borne Pathogens

Most seeds carry spores of various fungi, either on the

surface or within the tissues, and counts as high as 150,000 spores per tree seed have been reported. Some seedborne fungi can cause the death of seeds and seedlings whereas other fungi, for example species of Aspergillus, Penicillium, Chaetomium, Rhizopus, and Trichoderma, which are the genera most often isolated from seed samples of a wide range of species, are saprophytes. If improperly stored, the growth of saprophytic fungi on seed can drastically reduce viability, but with a few exceptions, e.g. *A. niger*, they are rarely implicated in causing the death of seedlings.

Eucalyptus Seed Mycoflora and Seed-Borne Pathogens

Lists of fungi isolated from samples of eucalyptus seed are often included with records from a range of other tropical tree species. Examples include Mohanan and Sharma, who also highlighted issues of seed collection processing, storage, seed health testing, treatment and certification and indicated where improvement was needed; Agmata, who provided the first list of tree seed-borne fungi in the Philippines; Mittal and others, who compiled a world checklist of microorganisms associated with tree seed; and Richardson who provided an annotated list of seed-borne diseases. Lists of seed mycoflora for Eucalyptus spp. alone are provided by Sharma and Mohanan, Tiwari and Sharma, and Reddy and others.

Several authors have demonstrated pathogenesis of seedborne fungi, either by isolation of putative pathogens from both seed and blighted seedlings, or by inoculation studies. Mittal and Mittal and Sharma studied the mycoflora of Eucalyptus hybrid (predominantly *E. tereticornis*) and *E. citriodora* (syn. *Corymbia citriodora*) and also the means of controlling pathogenic fungal species. Saxena detected 30 species of fungi on seed of *E. grandis* and related seedling mortality to seed mycoflora, and Michail and others reported on *fusarium postemergence* damping-off of *Eucalyptus* spp. and its control.

Four potentially pathogenic species, *A. niger, Fusarium sp., Penicillium canadense,* and *Rhizopus oryzae,* were found on seed of *E. citriodora.* Drake found that a species of Ramularia infected unopened capsules of *E. crebra* and *E. melanophloia* while still borne on the tree, and caused up to 50-percent sterility. Yuan and others inoculated *E. camaldulensis* seed with 14 putative pathogens isolated from seed and recorded seedling emergence compared to controls. All fungi except *Botrytis cinerea* and *Cytospora* sp. reduced emergence, with *Fusarium, Curvularia,* and *Pestalotiopsis* spp. being the most pathogenic.

In a later study with *E. pellita, A. niger, Dreschlera australiensis, Harknessia fumaginea,* and *Pestalotiopsis disseminata* all reduced germination and/or caused seedling mortality postemergence compared to controls. *Coniella australiensis,* a widespread leaf pathogen of eucalypts, was also isolated for the first time from seed. It was suggested that these fungi may have originated from leaf debris within the samples and that such tissue offers a possible means for dissemination of foliar and stem pathogens with seed.

Acacia Seed Mycoflora and Seedborne Pathogens

In a recent report on surveys of tropical acacia diseases by Old and others some reference is made to seedling diseases, notably those recorded in Thailand, but the emphasis of the report is on diseases of plantations and native stands. There are very few reports in the literature of seed-borne fungi causing diseases of acacia seedlings. Curvularia lunata has been recorded by Bhawani and Jamaluddin as causing shoot dieback of seedlings of *A. nilotica*. This fungus has been recorded previously from seed of *A. auriculiformis* and recently from *A. crassicarpa* and *A. aulacocarpa*.

Several of the pathogens recorded by Mohanan and Sharma as causing diseases of exotic acacias in India are putative seed-borne organisms. Yuan and others inoculated *A. auriculiformis* with the same 14 seed-borne fungi used to inoculate eucalypt seed, with similar results. *Fusarium* spp., *Dreschlera* sp., *Curvularia* spp., and *Pestalotiopis* spp. caused the most severe preemergence blight of inoculated seedlings.

The lists of pathogenic fungi on acacia trees in native stands and plantations in northern Australia and in India presented by Old and others and Sharma and Maria Florence include species of Pestalotiopsis, Phoma, Phomopsis, Lasiodiplodia, and Curvularia, which can be seedborne. The detailed etiology, however, of the diseases of acacias recorded in both reports is largely unknown. Old and others gave a brief account of a severe disease outbreak on *A. mangium* in 1992 in northern Australia caused by a *Cercospora* spp.

The pattern of disease incidence suggested that the origin of the outbreak was in the nursery where seedlings were raised, and plants grown from seed imported from Papua New Guinea initially showed the highest level of disease. Although severe damage occurred in several plantations, the disease was present at very low levels during the following year and was undetectable subsequently. Detailed analysis of all seed lots used in the plantings failed to isolate *Cercospora* sp. In view of several well-known examples of seed-borne *Cercospora* spp., this could have been the origin of the disease. Alternatively, the pathogen is indigenous and occurring undetected on native *Acacia* species in the region.

Risk from the Dissemination of Acacia Pathogens with Germplasm

As with eucalypts, the species recorded so far on the seed of tropical acacias are regarded as being already widely distributed throughout the regions where tropical acacias are grown. Also, they are generally pathogens with a broad spectrum of hosts. Unlike eucalypts, the seed of many acacia species are relatively large (10 mg to 1 g) and the presence of retained floral parts such as the aril offers niches for the carriage of saprophytic growth of facultative pathogens.

The recent report of an international workshop on diseases of tropical acacias gave an account of the status of a range of pathogens and the diseases they cause in Australia, India, and Southeast Asia. The stimulus for the workshop was the rapid expansion of plantations of tropical acacias, especially in Indonesia, Malaysia, and Thailand. Every year several tons of seed of *A. mangium*, the most widely grown species, are collected from native stands or seed orchards and dispatched between countries of the region.

There seems little doubt that this practice carries significant risk of transporting pathogens on seed or associated plant debris. The practice of immersing seeds in boiling water prior to sowing to break dormancy may reduce the likelihood of chance contaminants being spread, but there is an urgent need for more information on the diseases of these species and the possibilities of seed-borne spread.

Risks from the Dissemination of Pathogens with Casuarina Germplasm

One of the more significant diseases of *Casuarina equisetifolia* is the blister-bark pathogen *Trichosporum vesiculosum* (nom. illegit.). The status of this organism as a pathogen has been questioned. However, Narayanan and Sharma and Narayanan and others summarised the available information which strongly points to the fungus being a damaging pathogen of stressed trees. The fungus spreads by production of very large quantities of sooty spores within blisters under the outer bark, which rupture to release spores.

The etiology of the disease is imperfectly understood, but it is likely that wound infection is the major means of spread with further movement within an affected stand by root-to-root contact. Until 1994 the fungus was known as a pathogen of casuarina only in India and in Mauritius. However, blister-bark disease has been found in central Vietnam and was identified at two locations in Thailand in 1995 in replicate plantings of an international Casuarina provenance trial. The seed for this trial had originated in many countries, including India, and although there is no information as to the origin of the outbreak, it is possible that the fungus was seed borne.

T. vesiculosum would not be detected by standard seed testing methods; however, its fecundity suggests that chance contamination of seed could have been the origin of the simultaneous appearance of this disease in two locations in Thailand separated by approximately 200 km. There is a need for further research on this pathogen to determine its taxonomic affinities, its pathogenic status, and the etiology of blister-bark disease. Until further information is available it would be imprudent to disregard the possibility that the disease may be seed borne.

2

Evaluation of Integrated Plant Pathology

Pest-caused losses in the forest nursery may occur in several forms. The most obvious are the dead and dying seedlings observed in nursery beds. The economic loss represented by this seedling mortality may vary with the age of the affected seedlings. Seedlings that die early in the growing season usually represent less economic loss than seedlings that die later in the season, when greater investments have been made in them.

Pests also damage seedlings, making them unsuitable for planting. These seedlings are thrown away (culled) during lifting and packing. A third type of economic loss may occur when large numbers of seedlings are stunted or damaged during the growing season and require an additional growing season in the nursery. The expense of maintaining these seedlings in the nursery for the extra growing season is considered an additional economic loss attributable to the causal pest(s).

Nursery pests and their damage also cause economic losses in field planting. When seedlings are not available for planting during a particular season, investments in site preparation may be either partially or completely lost. The site may have to be prepared again when seedlings become available in subsequent years. Nursery pests originating in the nursery are frequently transported to field plantings on infected or infested seedlings. In the field, these pests may continue causing economic losses by killing, stunting, or deforming seedlings.

The most serious damage in field plantings involves the introduction of anew pest on infected or infested nursery stock and its subsequent spread to adjacent planted or natural stands. White pine blister rust and phytophthora root disease of Port-Orford cedar are examples of diseases that have been introduced into new localities on diseased nursery stock. That is why the transport of nursery stock between nurseries, States, regions, and countries should be carefully monitored. When such transport is necessary, seedlings should be carefully examined for pests. Most State plant quarantine agencies require seedlings to be examined and certified pest-free prior to interstate transport.

Pest Loss Evaluation

A pest loss evaluation program should be an integral part of any nursery operation. The program should include biological, economic, and environmental evaluations of pest-caused losses, along with pest management alternatives and post control appraisals. A first prerequisite in any nursery pest evaluation is identifying the pest problem and associated causal agent(s). Causal agents may include biological organisms such as insects and fungi, or a biotic factors such as the nursery environment and management's cultural practices. Additional background information that is helpful in planning nursery pest evaluations include:

1. Seedling species.
2. Seedling age when affected.

3. Portion of seedling affected.
4. Seedbed location.
5. Size and distribution of affected area.
6. Soil type and drainage.
7. Cultural practices.
8. Pesticide use.

Nursery pest evaluations may be designed either for general pest occurrence or for specific pest problems. The evaluation intensity may range from informal seedbed scouting to intensive randomised or systematic sampling procedures that permit statistical analyses. Preliminary sampling data may also be needed to determine the distribution and variation of the pest problem in order to select the most effective and efficient evaluation type and intensity. It is important to evaluate when the symptoms are readily visible.

Root diseases, stemrusts, and foliage blights, for example, are easiest to see late in the growing season. Proper timing of evaluation requires a knowledge and understanding of localised nursery pest problems. After pest damage is identified and quantified, cost-benefit and environmental impact analyses of pest management alternatives are needed to select the pest management strategy most economical and safest to the environment. Nursery pest management strategies include routine preventive practices as well as curative treatments after pest problems are detected, identified, and evaluated.

High product values and high pest hazards in nurseries frequently make preventive practices such as soil fumigation and chemical sprays desirable. However, the most effective and efficient pest management procedures involve an integrated nursery pest management strategy. The primary objectives of this strategy are to minimise potential pest losses and maximise production of high-quality, pest-free seedlings for improved survival and growth in field plantings. Pest evaluation of field plantings also is very important. Frequently, such evaluations are omitted orde-emphasised, or the results are not reported. Background information that is helpful in evaluating field planting problems includes:

1. Seedling storage (length, type, and container).
2. Site conditions (site index, soil type, and competition).
3. Planting conditions (moist or dry).
4. Planting method (hand or machine).
5. Seedling size (too small or too large).
6. Seedling root system (adequate, substandard, or needing additional root pruning).
7. Incidence of mycorrhizae on seedling feeder roots.

Field planting evaluations promote cooperative relationships with field foresters and relate the nursery manager's concern for successful forestation, which is everyone's primary goal.

Mycorrhizae

Specialised root-inhabiting fungi form beneficial associations with all forest tree species. These fungi invade the feeder root tissues and form modified roots called mycorrhizae (fungus-roots), which greatly increase efficiency of nutrient and water uptake. Most plants require mycorrhizae for normal growth and development in natural soils. Mycorrhizae can be classified into two primary types: ectomycorrhizae and endomycorrhizae.

With ectomycorrhizae, which are normally found on pine, spruce, fir, beech, eucalyptus, alder, oak, and hickory, the fungi grow between rather than within root cells. They form a structure known as the Hartig net between the cells, as well as a fungus mantle or cover on the surface of feeder roots.

The ectomycorrhizal feeder roots develop a swollen appearance, and in pines they normally have a forking habit. The endomycorrhizae are found on maple, sycamore, ash, gum, walnut, cypress, some poplars, and some other conifers. This mycorrhizal type also occurs on all agronomic crops, such as sorghum, corn, and grasses used as cover crops in tree nurseries. These fungi grow into the root cells, and a mantle or exterior cover is lacking.

Endomycorrhizal fungi frequently produce large, conspicuous, ornamented spores on hyphae attached to endomycorrhizal roots, which are detectable under low magnification and represent a reasonably reliable indicator of endomycorrhizal fungus root infection. Mycorrhizal feeder roots are responsible for the uptake of most of the nutrients and water from natural soils. These structures are essential to the health and vigor of plants, especially forest trees that grow in stressful environments.

Seedlings of ectomycorrhizal tree species commonly grown in fumigated nursery soils form ectomycorrhizae with naturally occurring fungi that originate from windblown spores. These spores are produced by mushrooms and/or puff balls in adjacent windbreaks, seedling beds, or forested stands. The lephora terrestris is the most common naturally occurring ectomycorrhizal fungus in bare root seedling nurseries in the United States. In southern pine nurseries, which usually produce

1-0 stock, 25 to 75 percent of feeder roots are ectomycorrhizal with *T. terrestris* by lifting time. This fungus is well adapted to the optimum growing conditions in nursery beds, but, unfortunately, notto the adverse conditions of many reforestation sites.

Hardwood seedlings grown in fumigated nursery soil may have poor endomycorrhizal development. Nonselective soil fumigation reduces populations of these fungi, which do not have the capability to rapidly recolonizethe partially sterilised soil, as doectomycorrhizal fungi. Endomycorrhizal fungi do not produce large, above ground fruiting bodies and windblown spores for effective recolonisation of soil. Because of inadequate endomycorrhizal development, these seedlings may grow poorly in the nursery.

Container nurseries have special problems. Seedlings are produced under artificial conditions, and artificial inoculations may be the sole source of mycorrhizal development. Thus, mycorrhizal inoculation in container nurseries is often required. Artificial seedbed and container inoculations with pure cultures of mycorrhizal fungi are likely to become standard practices in those operations where these fungi are absent or deficient.

Ectomycorrhizae

Recent research has identified an ectomycorrhizal fungus, Pisolithustinctorius (Pt), that is ideal for adverse planting sites. Procedures have been developed for using Pt in bare root and container nurseries. Studies have shown that when nursery seedlings are inoculated with this fungus, numbers of cull seedlings are reduced, and survival and growth infield plantings are increased. The increases have been most apparent on adverse forestation sites like strip mine spoil banks, but consistent and significant improvements have also been observed on many routine sites.

The tailoring of seedlings with Pt mycorrhizae before planting on adverse and many routine sites is certainly worthwhile. Use of the fungus may be generally beneficial on all sites because poor conditions, especially soil water deficits, during and following planting can make the best field sites temporarily adverse. Techniques for inoculation have been developed in research studies and tested in more than 100 nursery trials. Inoculation techniques are reliable, but they are not well understood by most nurserymen.

In bare root nurseries, Pt mycelial or spore inocula can be placed inprefumigated (preferably spring) seedbeds before or at seed sowing. Incooperation with private industry, the USDA Forest Service has developed an ectomycorrhizal fungus inoculum applicator to apply Pt mycelial inoculum during seed sowing in bare root nurseries. This machine uniformly applies bands of inoculum into the root zone and levels the soil in preparation for sowing seed in the nursery bed.

Mycelialinoculum can also be broadcast on the seedbed surface by hand, but using the machine reduces the inoculum needed by two-thirds and places it in the drill rows, where it is most needed. Spore inoculum can be applied as bulk and pelletised spores, on spore-encapsulated seeds, and with hydromulch applications. In container nurseries, Pt spore or mycelial inoculum can be added to the growing medium before containers are filled. Another technique is to apply either spore pellets or bulk spores onto the growing medium after seeds are sown. A third is to encapsulate the seeds in spores before they are sown.

In general, spore inoculum has not been as effective as mycelialinoculum in early establishment of Ptectomycorrhizae on seedling roots. In addition, encapsulation of seeds of some species with spores inhibits germination somewhat. Nevertheless, spore inoculum appears to be practical in some nursery operations. Inoculation failures in nursery trials are usually attributable to adverse soil conditions and cultural practices, and nonviable inoculum.

Certain pesticides, especially the systemic fungicide triadimefon (Bayleton®), inhibit formation of ectomycorrhizae from both Pt inoculum and natural sources. However, this fungicide is highly effective for fusi form rust control, and it is used routinely in nurseries where rust hazard is high. The main edaphic factor affecting ectomycorrhizae development is pH. Soil pH above 6.0 strongly inhibits development. Soil phosphorus up to 150 ppm and applications of nitrogen totaling 400 lb per acre during the growing season do not inhibit ectomycorrhizae.

Endomycorrhizae

Most major hardwood species form endomycorrhizae and require them for normal development in forest plantings. Adequate development of endomycorrhizae in the nursery has two benefits. First, seedling quality in the nursery is improved. Second, hardwood seedlings that have good lateral roots and good endomycorrhizal development before out planting do not die back as much after out planting as those with poor lateral roots and endomycorrhizal development. Hardwood seedlings with endomy-corrhizae are better able to compete with undesirable vegetation on the planting site.

The most important long-term effect is in phosphorus uptake. Forest soils usually do not have sufficient phosphorus in available form for optimum tree growth. Naturally occurring endomycorrhizae aid in phosphorus uptake. Without them, many trees would die. Fortunately, the endomycorrhizal fungi are present naturally in forest soils. Maintaining them in nursery soils requires special management.

Cover crops of such species as sorghum, corn, millet, and sudex are ideally suited for increasing the populations of endomycorrhizal fungi in soil. The crops also add organic matter to the soil and reduce water and wind erosion. In addition, the dense feeder root systems of most cover crops increase natural levels of inoculum of endomycorrhizal fungi. Unfortunately, soil fumigation can nullify all of these benefits if it is applied after the cover crop is worked into the soil and before hardwood seed sowing.

The nonselective fumigant removes most of the endomycorrhizal fungi from the top 8 inches of soil. Therefore, routine soil fumigation for hardwood seedling crops should be applied before the cover crop is planted. In addition, the cover crop should be permitted to mature. Maturity of annual plants appears to be essential for the completion of the life cycle of the endomycorrhizal fungi. Only at the end of the growth cycle are the spores for inoculating the tree crop produced in large quantities.

Some nurseries use two cover crop rotations in a single growing season in the hope of adding more organic matter to the soil. From the standpoint of endomycorrhizal development, this practice is not advisable if the second crop does not have adequate time to mature. Phosphorus fertilisation of nursery beds must be carefully monitored to permit good mycorrhizal development. Available soil phosphorus should be maintained between 75 and 100 parts per million (Bray II test). Less than 50ppm without mycorrhizae will inhibit seedling development, and higher than 100 ppm will inhibit endomy corrhizal development. Six to eight applications of nitrogen (50 lb per acre, as ammonium nitrate) are also recommended.

Seedlings with few mycorrhizal roots do not absorb water and nutrients from the soil efficiently. Thus, they suffer severely from transplanting shock and often die. Mycorrhizae are living, somewhat delicate structures on the feeder roots of plants. They can easily be damaged mechanically or by desiccation. Improper harvesting operations can strip seedlings of most of their mycorrhizae, as can careless handling, packing, storage, or shipping.

Excessive root pruning for convenience in packing, shipping, or planting can remove the majority of mycorrhizae as well as root tissues containing food reserves that will be needed during the hard times following out planting. Root systems must be kept cool and moist from the time the seedlings are lifted in the nursery until they are planted in the field.

Integrated Nursery Pest Management (INPM)

Integrated nursery pest management (INPM) is the most practical and ecologically sound approach for control of nursery pests because it is integrated with all other nursery management procedures. As a result, nursery managers are not limited to fighting damaging pest "fires" that should not have been ignited in the first place. INPM is accomplished through coordinated tactics that prevent pest damage from reaching economically damaging proportions while protecting the environment from potential hazards. The tactics may be classed as preventive, cultural, chemical, and biological. They are selected to achieve pest management goals in the most practical manner.

Pest Prevention Methods

The best kind of pest control is that which prevents pests from becoming a problem in the first place. Preventing problems, however, requires a great deal of knowledge and effort. For nurserymen, the key preventive activities are training; a combination of detection, diagnosis, and evaluation; and exclusion.

Training: Perhaps the most important part of a successful INPM program is the recurrent training of nursery personnel. Technology in applicable pest management techniques changes rapidly, and the nursery that does not keep abreast of the changes stands to lose a great deal.

Detection, Diagnosis, and Evaluation: Early pest detection, rapid diagnosis, and evaluation of either existing or potential pest problems are prerequisites to a successful INPM program. All nursery personnel should be aware of and alert to early indications of pest occurrence. Observations of abnormal seedling appearance should be reported immediately to appropriate pest specialists to facilitate rapid diagnosis and evaluation. Only in this way can the most effective, practical, and environmentally safe INPM practices be selected and applied before the pest problem becomes unmanageable.

Exclusion: Nurseries should strive to exclude pests that may be transported and spread in contaminated seedlings, soil, mulches, water, and equipment. Nursery sites with pest problems should be identified and recorded, and steps should be taken to minimise expansion. Pest exclusion procedures are probably the most effective, practical, economical, and safe INPM practices readily available for nursery application. The standardised systematic plant quarantine regulations established and administered by State agencies should also be followed. These practices are often the only effective way to prevent a nursery pest problem.

Cultural

Site Selection: When selecting sites for new nursery locations or for expansion, managers must give careful consideration to the type, texture, pH, and drainage of the soil; past cropping history and present indications of pest occurrence; and an adequate supply of uncontaminated water with low salinity and proper pH. The soil type for most tree species should be of coarse texture (primarily sand and silt, with a low clay content). The subsoil should be free of impermeable layers, particularly at shallow depths. Soil-borne pathogenic fungi, such as those causing damping-off and root rots, are generally least severe in coarse-textured soils with good drainage. Fine-textured clay soils with poor internal drainage are difficult to till and fumigate. The pH values of soil and irrigation water also influence the occurrence of several soil-borne pests. Soils recently in field crops often contain pathogenic fungi that attack nursery seedlings. Water can be contaminated with pathogenic fungi, nematodes, insects, weed seeds, salts, and herbicides detrimental to quality seedling production.

Species Requirements: The requirements of tree species vary, and the needs of the crop species should be identified and recorded. Pest problems are most severe when these requirements are not fully met in nursery beds. For example, site requirements, soil microbiological associations, mycorrhizal fungi, and cultural procedures differ between conifers and hardwoods. When they are produced in the same nursery, managers must adjust cultural practices and site conditions for the needs of each species.

Crop Rotation, Cover Crops, and Organic Matter Amendments: Crop rotation can often reduce pest problems. Continuous cropping of seedlings of a given species promotes the buildup of a variety of pests (pathogenic fungi, insects, nematodes, and weeds). Rotating susceptible and nonsusceptible seedling crops in the proper sequence will usually avoid significant damage. Cover crops are alternated with seedlings as a standard practice in most nurseries. These crops reduce wind and water erosion and provide soil organic matter. The various cover crop species differ in susceptibility to nursery pests.

Corn, peas, soybeans, and sorghum are susceptible to and promote the development of charcoal root rot and black root rot, while millet, sudan grass, and sudex are relatively resistant. After the cover crop is plowed under, it must be permitted to decompose before the soil is fumigated. When it is green, it may absorb or otherwise deactivate large quantities of methyl bromide and make fumigation ineffective. Organic matter improves tilth, nutrient and water retention, and aeration of nursery soils. It must be added to produce good-quality seedlings consistently.

Organic matter amendments may reduce nursery soil pest problems in various ways, such as promoting populations of beneficial soil fungus saprophytes and competitors and increasing biodegradation of residual toxic chemicals. However, the type and condition of the organic amendments must be carefully controlled. For example, fresh sawdust may reduce seedling growth because, as it decomposes, the microorganisms responsible for this process tie up available nitrogen. In addition, organic amendments such as pine needles, straw, and sewage sludge may harbor harmful pathogenic fungi, insects, nematodes, weed seeds, and chemicals.

Soil Acidity/Alkalinity: The degree of acidity (low pH) or alkalinity (high pH) in soil strongly influences the development of soil-borne pathogenic fungi. In general, damping-off and root rots are most severe in nursery soils with a pH above 5.5. In nursery beds and containers, soil pH levels above 6.0 have also significantly reduced the development of ectomycorrhizae on conifers. The pH can be lowered by applying elemental sulfur or ammonium sulfate. Lime is routinely used to increase pH in nursery soils. The pH of irrigation water can be lowered by metering prescribed amounts of sulfuric or phosphoric acid into the irrigation system.

Seedbed Sowing Date: In early spring, the cold, moist soil is conducive to the development and spread of certain damping-off fungi (Pythium and Phytophthora). Therefore, spring seeding should be delayed until the soil temperature approaches the optimum for seed germination and early seedling development. In the Deep South, excessively warm surface soil in late spring frequently causes "sun scald" and mortality on young, recently germinated seedlings. Consequently, either early spring or fall sowing is required for satisfactory seed germination and survival of young seedlings.

Seedling Density: Control of seedling density reduces the potential hazard for certain pests. At high densities, seedlings must compete for available nutrients and water, and growth and vigor are reduced. Higher seedling densities also promote denser foliage coverage, which reduces air movement and provides a more favorable microclimate for foliage disease fungi. High demands for tree seedlings for accelerated forestation programs have caused many nurseries to increase seedbed densities. This practice promises to be self-defeating.

Pest problems will increase, and the quality of seedlings will decline. As a result, huge investments in preparation of planting sites and in planting itself may be jeopardised to save a few dollars per thousand seedlings in the nursery. Nursery practices must be designed to produce the highest quality seedlings for successful forestation.

Mulches: Precautions are needed in the selection and utilisation of mulch materials to avoid the introduction and development of pest problems. As previously pointed out, pine needles, sawdust, and grain straw may contain pests such as pathogenic fungi and weed seeds. Such materials should be fumigated with a mixture of methyl bromide and chloropicrin- either 98 percent and 2 percent or 67 percent and 33 percent-at a dosage rate of 1.0 lb/yd3 of material.

In most cases, mulch materials are relatively dry and require additional water to provide adequate moisture for effective fumigation. In addition, a minimum 72-hour, and preferably longer, aeration period is required before fumigated mulches are used. These materials absorb large amounts of the fumigant, and long aeration periods are required to release these toxic chemicals.

Fertilisation: Fertiliser rates, composition, timing, and application methods can significantly affect pest populations, as well as those of associated beneficial soil organisms. Inadequate rates, incorrect formulations, and improper applications of fertilisers can stunt seedlings, burn foliage, and cause summer chlorosis and poor development of roots and mycorrhizae. Seedlings that are stressed in these ways are considerably more susceptible to a variety of soil and aboveground pests.

Excessive applications of nitrogen and phosphorus may increase damping-off in early spring and increase fusiform rust on slash and loblolly pines in the South. In addition, excessive applications of phosphorus inhibit formation of ectomycorrhizae on conifers and endomycorrhizae on hardwoods.

Irrigation: Inadequate, excessive, or improper irrigation may create nursery pest problems. As with inadequate fertilisation, inadequate water may cause seedling yellowing, stunting, and mortality -particularly during hot, dry weather. In addition, both conifer and hardwood seedlings growing with inadequate soil moisture and high temperatures (above 90 °F) are considerably more susceptible to charcoal or black root rots.

Excessive water may inhibit soil aeration, cause root mortality, and thereby increase seedling susceptibility to certain water-loving soil pests, such as Pythium, Phytophthora, and nematodes. Repeated excessive irrigation (particularly during late afternoon and evening) may increase the incidence of foliage disease (Botrytis in container nurseries) and decrease the development of mycorrhizae on both conifer and hardwood seedlings.

Temperature Controls: Temperature is one of the most difficult environmental factors to manipulate in bareroot nursery seedbeds. In greenhouse container nurseries, desired temperatures are most commonly obtained through combinations of heating, airconditioning, restricting sunlight (lathing), and irrigation. Both high and low soil and air temperatures significantly affect a number of pest problems, as well as seed germination and subsequent seedling growth. Temperatures in bareroot nurseries can be reduced with irrigation.

In addition, some species that are highly intolerant of high temperatures such as fir, spruce, and white pine, may require lath shading. The adverse effects of low soil temperatures in bareroot nursery seedbeds in early spring can be avoided by delaying the seedbed sowing date until the soil temperature is suitable for rapid seed germination and early seedling growth. However, as previously pointed out, this practice may have limited application in southern nurseries, where excessively high soil and air temperatures may frequently occur early in the growing season.

Soil Drainage and Subsoiling: Poor soil drainage invites problems with a variety of soil pests. Solutions include seedbed elevation, crowning, subsoiling, and the installation of drainage tiles. Sometimes surface water collects and stagnates in low spots in seedbeds. This problem can be eliminated by leveling and reshaping the seedbeds. Orienting seedbeds to promote maximum water runoff away from the beds will also decrease drainage problems and subsequent pest occurrence.

Subsoiling, a standard practice in most nurseries, is utilised primarily to break up underlying hardpans and to decrease the soil bulk density. Subsoiling increases soil drainage and aeration and permits deeper root penetration, particularly for hardwoods. Decreases in soil bulk density promote a more fibrous seedling root system with increased mycorrhizae development.

Eradicating Alternate Hosts: Southern fusiform rust and white pine blister rust are caused by obligate parasites that require specific alternate hosts (oaks and Ribes spp., respectively) to complete their life cycles. Where feasible and practical, the alternate hosts of these and other rusts should be eliminated within and around the nursery to minimise disease problems. Spores of these fungus pests can travel long distances, and complete eradication for extended distances from the nursery is frequently impractical. Nevertheless, alternate host eradication within and near the nursery seems well advised.

Windbreak Species: Tree windbreaks are quite useful, but tree species for windbreaks should be neither alternate hosts nor species susceptible to potential seedling pests. The use of similar windbreak and seedling crop species may supply fungus inoculum and insect pests that can readily spread to adjoining nursery seedlings. Existing windbreaks of susceptible species may require sanitation pruning or elimination.

Sanitation: Sanitation should be routinely employed to prevent the spread of pest problems within the nursery and to subsequent field plantings. Diseased or insect-infested seedlings should be rogued. Weeds and pest host-plants in and around the nursery should also be eliminated to reduce fungus inoculum, insects, and weed seed sources.

Seedling Grading and Culling: This practice may be required to minimise the transport of pest-infested seedlings to field planting sites. Seedling examinations during packing are highly beneficial in detecting inconspicuous root rots and other diseases, such as stem rusts and certain foliage diseases, that frequently are not detectable until late in the growing season. Also, diseases that are erratically distributed in nursery seedbeds are considerably more difficult to eliminate in bed culling than in seedling culling during packing. Managers should carefully evaluate the present trend to de-emphasise seedling grading and culling during the packing operation. The benefits of this practice include improved seedling quality and reduced pest damage potential in field plantings.

Chemical Treatments

Chemical treatments are integral components of INPM, and a variety of chemicals are needed for use before and after seed sowing. However, chemicals should be used only (1) in combination with other INPM practices, where they have failed to produce satisfactory results; or (2) in situations where other INPM practices are not available.

Soil Fumigation: Soil fumigation before seed sowing is the most effective means of reducing soil-borne pathogenic fungi, insects, nematodes, and seeds of some weed species. Although several soil fumigants are registered for use in forest tree nurseries, the most effective are the methyl bromide-chloropicrin formulations. The formulation of 67 percent methyl bromide and 33 percent chloropicrin is most effective in reducing soil-borne fungi that have resistant spore stages, such as those that cause charcoal, black, and Cylindrocladium root rots on conifers and hardwoods.

The formulation of 98 percent methyl bromide and 2 percent chloropicrin is widely used where a broader spectrum soil fumigant is needed for a variety of soil pests. Soil fumigation is relatively expensive, and there is no excuse for doing it poorly.

Soil fumigants are nonselective in activity. Thus, non-target beneficial organisms, such as the mycorrhizal fungi, are reduced. The ectomycorrhizal fungi associated with most conifer species and other beneficial microorganisms necessary in a suitable soil environment usually re-infest fumigated soil fairly rapidly by airborne spores blown in from adjacent soils. Re-infestation may be delayed where adjacent soils do not provide sufficient inoculum. A much more drastic negative effect of soil fumigation may be observed on certain hardwood species, such as sweetgum, yellow-poplar, and black walnut, which have endomycorrhizae.

The endomycorrhizal fungi produce fewer spores, and they produce them below ground. These spores are not transported by air currents. Consequently, re-infestation of fumigated soil takes a much longer time. It is possible that an effective soil fumigation may minimise one problem (a soil-borne pathogen) and create another (reduced mycorrhizal fungi and other beneficial microorganisms) of equal or greater importance.

Soil Pesticide Applications: Pesticides occasionally must be applied to the soil to control certain diseases, nematodes, insects, and weeds. Fungicides that are registered for use in forest tree nurseries are limited in number. They range from broad-spectrum compounds, such as captan, to those with very specific range of activity, such as metalaxyl (which affects only species of Phytophthora and Pythium). Generally, these compounds are applied after tree seeds germinate and a specific problem has been identified. They are applied either as soil drenches or as low-volume foliar sprays, followed by sufficient irrigation to move the compound down into the root zone. Since a fungicide is usually applied to soil in response to an existing disease problem, multiple applications are often required to maintain control of the problem.

Nematodes rarely cause problems in forest tree nurseries, but where they do, a nematicide should be applied before sowing. Granular formulations must be incorporated into the soil; liquid formulations can be injected into the soil at a prescribed depth by using chisel plows. Soil insects also rarely cause problems in forest tree nurseries. When an insect problem is identified, an appropriate insecticide should be applied in a water solution of sufficient volume to move the insecticide to the target site.

Special precautions are warranted when prescribing "protective" insecticide treatments involving both target pest and nontarget biocontrol organisms (e.g., spider mites and associated insect predators) on the same host seedling species. Indiscriminate or improper use of insecticides in such cases could be more detrimental than beneficial. Significant advances in nursery weed control have been made in recent years. Most nursery managers routinely apply one or more herbicides to the soil surface at planting time. These preemergence herbicides prevent germination of various weed species. Other post emergence herbicides are registered for controlling weed species that do germinate. Conscientious and safe use of registered herbicides in forest tree nurseries has virtually eliminated hand-weeding.

Foliage Pesticide Applications: Pesticides often must be applied with hydraulic spray equipment to foliage in forest tree nurseries. Fungicides and insecticides may be either contact or systemic. Contact fungicides must be present on the susceptible plant part to be effective; these compounds have little or no effect against established infections. Since contact fungicides do not move on the plant surface or inside the plant, they must be applied frequently to protect newly formed tissues. Systemic fungicides enter a plant's vascular system and are translocated throughout the plant. Because of this unique characteristic, systemic fungicides require fewer applications to protect all tissues.

In addition, many systemic fungicides exhibit therapeutic (curative) activity against infections already established within the plant. One notable example is triadimefon (Bayleton®), a compound used to control fusiform rust in southern nurseries. Triadimefon applied as a foliar spray controls fusiform-rust infections that occur within 7 days prior to treatment, and it provides protection for at least 21 days after application. Thus, three equally spaced applications of triadimefon between mid-April and mid-June are as effective in controlling fusiform rust as 30 to 40 applications of the contact fungicide ferbam.

Foliar herbicides used in forest tree nurseries must be toxic to target pests (weeds and grasses) but nontoxic to the tree seedling crop species. Appropriate herbicides for specific weed problems and conditions in forest tree nurseries are available.

Seed Treatment Chemicals: Chemicals are applied to forest tree seeds to protect them against birds, animals, and seed-borne pathogenic fungi. They also help to prevent soilborne damping-off organisms from attacking emerging seedlings. A seed treatment has recently been developed to protect emerging southern pine seedlings from fusiform rust. The most commonly used seed treatment chemical is thiram. This chemical is both a bird or bird and animal repellent, and a fungicide that has activity against certain seed-borne and soil-borne pathogens.

There are presently no systemic seed treatment compounds registered for use against seed– borne pathogens. The one systemic fungicide registered for use as a preplant seed treatment is triadimefon. Treating loblolly and slash pine seeds either in a solution of triadimefon (10 oz. a.i. per 100 gal of water) for 24 hr or a dry product mixture at the same dosage rate and exposure period protects newly emerged seedlings from fusiform rust for at least 14 days. As a result, foliar sprays with this fungicide can be delayed until seed germination and seedling emergence are fairly complete.

Biological Techniques

Although biological techniques are the most desirable for pest management, effective and practical applications in nurseries have been rather limited. One of the most successful examples has been the recent application of selected ectomycorrhizal and endomycorrhizal fungi in conifer and hardwood nurseries, respectively. Other biological controls include resistance, parasites and predators, and competitive and saprophytic fungi and bacteria.

One of the most significant improvements in management of forest tree nurseries involves seed-source selection. In the early days, nursery personnel were required to collect their own seeds. They often did so with little regard for tree form, growth rate, or disease resistance. In many cases, inferior seedlings were the result. Most forest tree nurseries now use seeds from carefully managed seed orchards. These orchards are comprised of clones selected for desirable tree growth and pest resistance.

Seed orchard clones are meticulously selected from field sources, and their subsequent progeny are carefully evaluated on a variety of field planting sites. The clones with poorly performing progeny are quickly rogued from the orchard. An effective, efficient, and rapid technique for testing selected seed orchard pine clones for fusiform-rust resistance is presently provided by the Resistance Screening Center, Asheville, NC 28806. These tests are conducted under highly controlled conditions, and the results are available considerably sooner than those from routine field progeny tests.

At present, a large number of forestry agencies are establishing third-generation seed orchards. Some seed orchards producing disease-resistant material have also been recently established. Consequently, seed sources for future use in forest tree nurseries will be even more resistant to pests. Biological agents can significantly reduce pest populations (particularly insects) in the nursery. These agents are present in nature, and their development usually coincides with the development of pest populations.

Unfortunately, the introduction of parasites and predators as biological control agents has had only limited success in nurseries. The majority of the fungi and bacteria in the nursery are beneficial rather than harmful. In fact, certain types of fungi and bacteria are required for successful production of tree seedlings and cover crops. Most soil fungi and bacteria are saprophytic on soil organic matter. Some fix nitrogen; some are antagonistic to or compete with soilborne pathogenic fungi.

Insufficient populations of these beneficial soil organisms greatly impede the processes of organic-matter decomposition and nutrient fixation. Most of these fungi and bacteria are the initial colonisers of recently fumigated soil. Certain fungi such as the Trichoderma species are highly antagonistic to, and competitive with, pathogenic fungi.

3

Diagnosis of Pesticides in Seed Sector

Many disease, insect, and weed problems are inherently associated with specific soil properties. Thus, in addition to affecting many cultural practices, soil properties frequently dictate the timing, method, and effectiveness of pest management practices.

Texture

Texture (size distribution of soil particles) is the most important soil property affecting occurrence, severity, and management of seedling diseases and insects. The ideal soil texture for most nurseries is either a sandy loam or a loamy sand that contains a maximum of 15 to 20 percent silt-plus-clay particles. An example of the effect of soil texture occurs with preemergence damping-off. This disease is less severe in coarse-textured than in fine-textured soils because the slower rate of seedling emergence in fine-textured soil extends the time seedlings are exposed to fungus pathogens.

On the other hand, corky root disease of Douglas-fir seedlings is more severe on coarse-textured soils. Both the pathogenic nematode Xiphinema bakeri and the corky root disease occur only in nurseries with coarse-textured sandy soils that have pore spaces large enough to allow movement of X. bakeri. In addition, certain soil insects, such as the black vine and strawberry weevils and white grubs, are favoured by coarse-textured soils.

Soil texture also affects the type and effectiveness of pest management practices. In general, pest management is more effective on coarse-textured sandy soils because it is easier to control soil temperature and moisture conditions. In addition, soil fumigation is more effective in coarse-textured soils because the fumigant moves more easily through the larger pores.

Soil Moisture

Soil moisture is a major factor affecting the incidence, development, and control of nursery pest problems. Ideally, nursery soils should be well drained; excess moisture-rather than moisture deficiency-is the most common problem. Soil moisture deficiency is rare in modern nurseries equipped with adequate irrigation systems. Soil-borne pests favoured by excess moisture include damping-off and root rots caused by fungi in the genera Pythium and Phytophthora. Larvae of the marsh crane fly, Tipula paludosa, are also more likely to occur and damage seedlings in soils with excessive moisture.

Other types of damping-off and root disease fungi such as Rhizoctonia are favoured by moistured efficient soils, as are Phomopsis canker on Douglas-fir and Cytospora canker on hardwoods. The success of many pest management practices is highly dependent upon soil moisture levels. For example, the effectiveness of soil fumigation is frequently determined by soil moisture conditions during and following application.

Soil Temperature

Soil temperature affects seed germination, seedling growth, and pest development and survival. Soil temperature is determined primarily by climate, but soil moisture and colour also affect it. Dark-coloured soils absorb heat and develop higher temperatures than light-coloured soils. Irrigation is one of the most commonly used practices for reducing soil and seedling temperatures.

Disease development can be greatly influenced by temperature. Preemergence damping-off, for example, caused by soil and seed-borne fungi such as Fusarium spp., is usually most severe during and after extended periods of cool, moist weather. This type of weather also impedes development of woody tissues resistant to the post-emergence form of this disease. Conversely, Fusarium root rot of Douglas-fir seedlings, charcoal root disease, and black root rot of pine are favoured by warm soil temperatures.

Likewise, increases in populations of the soil-borne pathogenic nematode X. bakeri coincide with seasonal fluctuation in soil temperatures. However, insect pests in nursery soils appear to be considerably less affected by short-term alterations in soil temperature than are diseases. The white grubs are the exceptions; the length of their life cycle is affected by prevailing soil temperatures in different climatic zones.

Soil Reaction (pH)

Most conifer and hardwood seedlings grow best in slightly acid soil, with a pH of 5 to 6. Soil-borne disease problems are most severe at pH 6 and above. Reducing soil pH is one way to minimise diseases without negative effects on seedling growth and quality. Damping-off is probably the best example of a seedling disease that is affected by soil pH. Acidifying the soil to control this disease has been used for many years and remains an effective and practical cultural practice, especially for alkaline soils.

However, before acidifying the soil, make sure that certain acid-tolerant pathogens, *Cylindrocladium* spp. for example, are not present. If they are present, lowering the soil pH may increase disease severity. Several damping-off fungi also cause root rots, and lowering the soil pH can be helpful in controlling them. Insects apparently are relatively insensitive to soil acidity within the range occurring in most nurseries.

Nutrients

The amount and type of nutrients can affect several diseases. For example, fusiform rust of loblolly and slash pine seedlings is significantly affected by soil fertility. Both host susceptibility to the disease and subsequent damage increase with nitrogen and phosphorus fertilisation. This rust fungus appears to infect vigorous, fast-growing trees more readily than slow-growing ones.

Calcium and phosphorus deficiencies contribute to higher damping-off losses, while excess nitrogen results in greater disease severity. The calcium and phosphorus deficiencies probably delay the formation of seedling woody tissues resistant to damping-off. Increased nitrogen most likely promotes formation of succulent, susceptible tissues.

Soil fertility is also important in at least one nematode-caused seedling disease: corky root disease caused by X. bakeri. A comparison of the soil nutrient levels within disease-free and infested areas suggests that disease severity is correlated with low soil fertility. Soil fertility apparently has little effect on nursery insects.

Soil Organic Matter

Addition of soil organic matter required to maintain proper nutrient and moisture-holding capacities and aeration usually does not affect soilborne nursery pests. Negative effects of organic matter soil amendments can generally be traced to using too much or the wrong types of organic matter. In one such case, increased damping-off was observed following the use of buckwheat.

Other types of organic amendments, such as sawdust, may provide fungi such as *Fusarium* spp. and *Cylindrocladium* spp. with food bases for survival. Addition of fresh, nondecayed sawdust may also promote nitrogen deficiencies (yellowing and stunting) of seedlings. This condition is primarily caused by the increased nutrient requirements of the large numbers of soil organisms (primarily bacteria) involved in the decomposition of the sawdust.

Diagnosis of Pest Problems

The first step in diagnosing a pest problem is recognising that a problem exists. To do this, managers must know what healthy plants look like under all conditions. Once a healthy standard has been established, they can spot abnormalities. The next step is determining the cause. Often, a systematic elimination of possible causes is more helpful than trying to select a single causal factor out of many possibilities.

Determining the cause of a plant abnormality requires a thorough examination of both individual seedlings and the nursery in general. The nursery manager should consider location of the beds within the nursery, along with information on cultural practices such as root pruning, fertilisation, and pesticide application in the initial phase of the diagnostic process. It is particularly helpful if the manager can examine both affected and non-affected areas because comparing differences between them may lead to solving the problem.

The pattern of occurrence of symptoms and signs is important in diagnosis. Pockets of affected seedlings with a progression of symptoms from the centre outward, for example, may indicate disease or insect infestation that is spreading. Seedling damage in depressions may indicate problems such as chemical buildup, poor drainage, or pests that are favoured by poor drainage. Perhaps the most important part of the evaluation is examining the individual affected seedlings. Look carefully at foliage, stems, and roots of seedlings.

The foliage is the best place to begin the examination because it is the first and most obvious portion of the seedling to show visible effects of abnormal conditions. Most foliage diseases caused by fungi, such as needle blight of conifers, and leaf spot, rust, and anthracnose on hardwoods, are characterised by small, discrete, usually darkened necrotic areas, often with fruiting bodies of the causal fungus developing on the dead tissue. Specialists use these fruiting bodies in identifying the causal agent and prescribing controls.

Some fungi that cause needle spots on conifers, such as Lophodermium spp., do not form fruiting bodies until most or all of the infected needle dies. The presence of needle spots on pines on which fruiting bodies never develop may indicate needle infection by stem rusts. Other foliage symptoms, such as general chlorosis and needle tip necrosis on conifers, or large irregular necrotic blotches and marginal necrosis on hardwoods, are more difficult to diagnose because they may be caused by air pollution, unfavourable environmental or soil factors, or root infection.

Insect damage is considerably more obvious, due to the actual presence of the pest or conspicuous damage to the foliage. The foliage may be chewed, yellowed, or otherwise deformed by insect feeding. If a problem is found on the foliage, proceed to a close examination of seedling stems. Death or discoloration of the distal portion of the seedling or branch, with the remainder of the seedling appearing healthy, is a good indication of fungus, insect, or animal attack on the stem.

Infection by pathogenic fungi is characterised by discrete necrotic areas, usually with a sharp line dividing the healthy and infected tissues. Fruiting bodies of the causal fungi are often present. Sunscald or chemical burns may mimic damage by pathogens, but the lesions of the former are often bleached rather than dark. Insect and animal attack are usually readily identifiable by the feeding and gnawing wounds. Seedling tops and branch tips killed by pathogenic fungi are usually brick red, whereas those killed by insect or animal girdling are straw yellow. Stem galls on pine seedlings are a good indication of stem rust infection.

But galls on hardwood seedlings are likely to be caused by insects. If girdling by pathogenic fungi, insects, or animals occurs low enough on the stem, the entire top of the seedling will die. In these cases it is important to examine such seedlings as soon as they exhibit the first symptoms of chlorosis or wilting, when the root system may still be alive and the possibility of root problems can be more easily excluded. Death of the entire top of the seedling usually indicates the presence of a root problem.

Root-related problems are often the most difficult to diagnose because both a biotic and biotic factors can result in similar symptoms. For example, excessive soil moisture can result in symptoms similar to those from fungus or nematode attack. However, when seedlings exhibit symptoms such as root blackening, lesions, shedding of the root cortex, orroughening of the bark on larger roots in the absence of obvious adverse soil factors, you should suspect fungus ornematode attack. Diagnosis of these problems usually requires laboratory analysis of plant tissue and/or soil because symptoms of most root diseases are similar and the causal agent is usually not evident from routine examination. The causal agent must be identified before proper control recommendations can be made.

Pesticides Regulation

Federal regulation of pesticides is found primarily in three acts-the Federal Insecticide, Fungicide, and Rodenticide Act of 1972 as amended; the Resource Conservation and Recovery Act of 1976; and the Comprehensive Environmental Response, Compensation, and Liability Act of 1980.

Federal Insecticide, Fungicide, and Rodenticide Act (FIFRA)

This law defines the conditions for developing, registering, and using pesticides. It has been amended several times, with the 1978 amendment producing significant changes in the requirements for registration, classification, and use of pesticides. The requirements for registration mainly affect pesticide manufacturing companies and will not be addressed here. The 1978 amendment gave States broader authority and responsibility for registering pesticides. States automatically have authority to register pesticides for use within the State for "special local needs."

Formerly, States had registration authority only with approval of the Environmental Protection Agency (EPA). Now, however, EPA may disapprove State registration if the use differs from Federal registration, if it creates an imminent health hazard, or if the State has authorised the use on crops where adequate crop tolerances or residue levels have not been established by EPA. Two other separate, but closely related, areas in FIFRA have also been amended-classification of pesticides and applicator certification.

To make maximum use of the State programs for training and certification of applicators of restricted-use pesticides, EPA is authorised to classify pesticide uses by regulation as a separate part of the registration process. Designation of restricted-use pesticides supports State applicator certification programmes, in which almost all States train and certify applicators of restricted-use pesticides. Restriction of use also provides EPA an alternative way to reduce pesticide risks besides outright cancellation of their registration. The provisions of FIFRA are legally binding. Violations of the provisions in the act carry both civil and criminal penalties. Private applicators who violate the act are subject to civil or criminal penalties of up to $1,000.

Commercial applicators are liable for civil penalties up to $5,000 and criminal penalties of up to $25,000 and 1 year in prison. FIFRA explicitly gives States the primary responsibility for enforcing requirements of the act. FIFRA also requires that commercial applicators keep records on their employment of restricted-use pesticides. The content and length of time the application records must be kept vary from State to State, but 2 years is generally the maximum that records must be kept. State enforcement agencies have the authority to make unannounced visits during normal business hours to inspect the records of commercial applicators.

The 1978 amendments also clarified an issue of broad concern-the legality of pesticide uses or practices not addressed in the label direction. The following pesticide-use practices have been specifically excluded from the definition of "use inconsistent with the label" and are therefore permissible:

1. Application of the pesticide at less than the labelled dosage, concentration, or frequency.
2. Application of the pesticide to control an unnamed target pest, as long as the crop, animal, or site is included on the label.
3. Pesticide applied using methods not specifically prohibited by the label wording.
4. Application of pesticide(s) with fertilisers, if not prohibited by the label.

Resource Conservation and Recovery Act (RCRA)

The second Federal regulation governing pesticide use is the Resource Conservation and Recovery Act (RCRA) of 1976. RCRA is designed to extensively monitor hazardous wastes. The act identifies solid wastes that are hazardous and sets forth requirements that govern their handling, storage, and disposal. This law applies to all hazardous wastes unless they are specifically exempted. The act is administered by the EPA, with assistance from the Materials Transportation Bureau of the U.S. Department of Transportation. The major provisions of the act for controlling hazardous wastes include:

1. A definition of "hazardous waste."

2. A manifest system to monitor hazardous waste from its generation to its disposal. This system is frequently called "cradle to grave tracking."
3. A permit requirement for facilities that treat, store, or dispose of hazardous waste.
4. A requirement that every State must have a hazardous waste programme.

Any installation that stores, disposes of, transports, or offers to transport hazardous waste must obtain an EPA identification number from the Regional EPA Administrator. Once the "waste" designation has been applied to any pesticide on EPA's hazardous waste list or residues of these pesticides resulting from a spill cleanup, they can be stored for only 90 days before disposal. Otherwise, a storage permit must be obtained from the Regional EPA Administrator.

Specific pesticide waste regulations were developed by EPA in the Code of Federal Regulations (CFR 260-266 and 122-124) and became effective in October 1980. Pesticides listed in subpart D of Part 261 of the act are not classified as hazardous wastes until the decision is made to dispose of them or they are stored pending disposal. Pesticides that have been determined to be excess are still considered pesticides. Pesticide containers that have been triple rinsed are not considered hazardous wastes under these regulations and can be disposed of either at approved landfills or by burial.

Comprehensive Environmental Response, Compensation, and Liability Act (CERCLA)

CERCLA established what is known as the "Superfund" for cleaning toxic wastes. This law was enacted in 1980 and is coded in Title 42 of the United States code, beginning in section 96D. The Superfund, initially $1.6 billion, was set up to pay for cleaning up spills of hazardous materials and the disposal areas themselves. The act specifies who is liable for reimbursing the fund. It also requires that spills be reported. There are some basic differences between RCRA and CERCLA. RCRA is not self-implementing. Rather than telling companies what they must do, Congress directed EPA to formulate regulations that would control company activities.

CERCLA's approach is very different: it establishes liabilities and obligations and does not require promulgation of regulations to be effective. A second important difference between the two bills relates to the time periods they are designed to control. Whereas CERCLA is designed primarily to determine liability arising out of historic waste management practices, RCRA is designed to affect current hazardous waste management activities. Another difference is that while the principal burden of implementing CERCLA rests with the Federal Government, the requirements of RCRA are primarily a State responsibility, subject to EPA approval.

Nurseries are affected by these laws when they dispose of pesticides that are classified as hazardous (RCRA), or when problems arise (CERCLA). Nursery managers who have questions regarding this subject should get in touch with the hazardous waste management officials in their State. Pesticides may be used safely and effectively to control nursery pests, but careful attention must be paid to both the Federal and State laws governing their use. Anyone having questions regarding the legal use of pesticides should contact the proper pesticide control official.

State Pesticide Regulations

In addition to the three Federal laws, States also have

regulations governing the purchase, use, and disposal of pesticides. Before any pesticide use project is initiated in any State, a responsible official must determine that all applicable State pesticide regulations are being followed.

4

Applied Plant Pathology

Disease survey is an important component in the management of crop diseases. Regular surveillance of plant diseases has to be undertaken to obtain details on the intensity and incidence of the diseases and weather factors like temperature, relative humidity etc., and to forecast the plant diseases.

Survival of Pathogens

Any pathogen can cause disease under favourable conditions. The only requisite factor is that the pathogen must come in contact with the host for the development of the disease. Pathogen itself or its parts that are capable of causing disease when brought near a host is called inoculum. Fungal pathogens are diversified where the vegetative body (hyphae), dormant mycelium, (embedded in the embryo of seeds or other plant parts), special reproductive structures (rhizomorphs, sclerotia, chlamydospores), various types of asexual spores (sporangia, sporangiospores, zoospores, conidia) and sexual spores (oospores, zygospores, resting spores, ascospores, basidiospores etc.), serve as inocula.

In the case of viruses and plant pathogenic bacteria, the individuals are acting as inocula, since they do not produce any special type of infective units like resting spores or endospores etc. But in the case of Streptomyces sp. (Actinomycetes), fragments of filaments and spore-like cells serve as inocula. In phanerogamic parasites, seeds are the potential inocula. Seeds in the soil survive for longer period. Orobanche seeds survive for about 13 years.

Seeds are abundantly produced for their multiplication which could attack the host plants. But dodder is an exception because broken bits of shoot can attack host plant. In any locality a time lag exists between harvest of a crop and subsequent sowing. Year after year, diseases appear in the newly sown crops. There should be some link between the previous crop and the subsequent new crop to revive or continue the life cycle. The existence of the pathogen between the two crop seasons is the vulnerable period in its life cycle. Hence knowledge of the survival of the pathogen in the off-season is useful for the plant pathologists to device effective control measures.

The establishment of a plant pathogen in a geographical location presupposes its ability to survive, not only during its parasitic relations with its hosts, but also during off-seasons in which the hosts are not growing. In temperate zones, plant pathogens must be adapted for survival overwinters or oversummers, like the powdery mildew pathogen that attacks fall-seeded wheat. In the Torrid zone, plant pathogens must be able to survive the dry seasons, during which susceptible plants are not growing. These sources of survival of pathogens or the sources for renewal of infection chain can be grouped as follows:

1. Survival by means of specialised resting structures
2. Survival as saprophytes

3. Survival in vital association with living plants
4. Survival in association with insects

Structures of Plant Pathogens

Enduring structures of plant pathogens may be as simple as conidia or as complex as perithecia. Apparently, ascospores or conidia derived from them, serve to carry the pathogen causing peach-leaf curl (Taphrina deformans) over the winter. Conidia of Alternaria solani, the pathogen of early blight of potato and tomato, survive for eighteen months in dried diseased leaves. Specialised thick-walled chlamydospores of Fusarium and other Imperfect fungi, spores of many smut fungi and the amphiospores, uredospores and teliospores of certain rust fungi also are important enduring structures.

The resting spores of Plasmodiophora brassicae may survive for ten years in soils infested upon the disintegration of clubbed roots. The oospores of downymildew fungi survive in the soil between growing seasons. In fact, oospores of the fungus that causes onion mildew do not germinate until several years after their formation. Some fungi survive unfavourable seasons in the form of sclerotia. Those produced by the omnivorous cottony-rot fungus, Sclerotinia sclerotiorum, can survive for years in a dry atmosphere. They decay rapidly, however, in warm moist soil.

Cold induced dormancy probably accounts for their ability to survive winters in temperate zones. Some powdery mildew fungi and other ascomycetes survive with plant refuse. Parasitic phanerogams survive in the form of seeds and as eggs, cysts and larvae of parasitic nematodes serve as overseasoning structures.

Saprophytes

The ability to live saprophytically enables many plant

pathogens to survive in the absence of growing susceptible plants. Saprophytic survival usually occurs in or on the soil. Waksman distinguished between soil inhabitants and soil invaders; the former comprise the basic fungal flora of the soil, whereas the later are short lived exotics. As applied to the root infecting fungi soil inhabitants are unspecialised parasites with a wide host range that are able to survive indefinitely in the soil as saprophytes; soil invaders (root inhabiting fungi) are more specialised parasites that survive in soils in close association with their hosts.

Most plant pathogenic fungi and bacteria are soil invaders, but some pathogens, notably Rhizoctonia solani and Pythium debaryanum which cause seedling blights and root rots, live saprophytically in the soils. The microbiological balance in the soil markedly affects the survival of saprophytic plant pathogens there. Apparently, Sanford was the first to suggest that control of potato scab by green manuring with grass might be due to the antagonistic action of saprophytic organisms flourishing on the green manure.

Not only do soil saprophytes antagonise other microorganisms by toxic action, but some such as Trichoderma lignorum actually parasitise Rhizoctonia solani and other soil-borne pathogens. Despite antagonism and parasitism by other organisms, many plant pathogens survive in the soil as inhabitants or invaders. The special conditions that favour biological control of plant pathogens in sterilised soil or in culture are nonexistent in field soil, in which there is a complex microflora and a low concentration of nutrients.

Certain plant pathogens survive between growing seasons as saprophytes in the dead tissues of susceptible plants. Such organisms are only incidentally associated with the soil, and live only as long as tissues of susceptible plants are available to supply nutrients. Most plant-pathogenic bacteria and many specialised parasitic fungi survive in this manner. The apple scab pathogen (Venturia inaequalis) lives parasitically in leaves and fruits during the growing season, but becomes saprophytic in fallen leaves.

Perithecia form in these leaves during the winter, but ascospores do not form until spring. Ascospores of certain other ascomycetes mature during the winter, but are protected from adverse conditions by perithecial walls. Soil inhabitants include obligate saprophytes and facultative parasites and they are exo-pathogens. Whereas soil invaders (root inhabitants) include facultative saprophytes and obligate parasites and they are endopathogens (root infecting fungi).

Plant pathogenic bacteria can saprophytically survive or actively multiply in the rhizosphere or rhizoplane of healthy host and non-host plants. Erwinia carotovora subsp. carotovora has been considered to survive in soil. However, some recent studies have shown that soft rot Erwinia cannot persist for a long time in fallow soil. E.carotovora subsp. carotovora multiplies in the rhizosphere of many cruciferous plant species, where the population can readily increase from 102 cells /g in fallow soil to 104 to 106 cells /g in soil subjected to the rhizosphere effect of chinese cabbage.

Pseudomonas glumae, the causal agent of bacterial grain rot of rice, remains on rhizosphere and/or rhizoplane of the rice plant from germination to tillering stage. Burkholderia solanacearum and species of Agrobacterium are best known with a prolonged soil phase which can be regarded as the true soil-borne pathogens.

Vital Association with Living Plants

Survival of the plant pathogens in vital association with living plants is grouped into:

Seed: The pathogen of loose smut of wheat, Ustilago nuda tritici, enters the stigma and style and infects the young seed, in which it survives as mycelium. The seed-infecting pathogens that cause loose smut of wheat and loose smut of barley are strikingly different from other smut fungi that attack cereal crops..Most of the others survives from season to season either in nonpathogenic association with seed or as spores in the soil. Colletotrichum lindemuthianum, the causative organism of bean anthracnose, can also infect the seed; unless the seed is killed, new infections are initiated by the fungus in newly sprouted bean seedlings. The bacteria that cause bean blights and bacterial blight of cotton survive the winter in infected seed.

In Mexico, the fungus of late blight of potatoes (Phytophthora infestans) produces oospores but in colder regions of the world, the fungus overwinters as mycelium in diseased tubers. Seed may be externally infected or internally infected by plant pathogenic bacteria during the course of development and maturation in fruit or pod. Most seed-borne bacteria survive as long as the seed remains viable; P.syringae pv tomato has been shown to survive in dried tomato seed for 20 years, X.campestris pv. phaseoli and Curtobacterium flaccumfaciens pv. flaccumfaciens in bean seed for 15 years, and P.syringae pv. phaseolicola in bean seed for 10 years.

Many plant pathogenic bacteria also survive in or on vegetatively propagated planting materials such as bulbs, corms, tubers, rhizomes or cuttings. Corynebacterium michiganensis subsp. sepedonicus and Burkholderia solanacearum may be present in the vascular system of potato tubers. Lenticels of potato tubers may carry soft rot Erwinia at the maximum level of 100 cells/lenticel, although the infested tubers do not necessarily develop soft rot in the field.

Collateral hosts: Collateral hosts are those which are susceptible to the plant pathogens of crop plants and provide adequate facilities for their growth and reproduction of these pathogens. Weeds which survive and live during non-cropping season provide for the continuous growth and multiplication of the pathogen. For example, the fungal pathogen for blast disease of rice, Pyricularia oryzae can infect the grass weeds like Bracguarua mutica, Dinebra retroflexa, Leersia hexandra, Panicum repens, etc., and survive during off-season of rice-crop.

As soon as a fresh rice crop is raised, the conidia (inoculum) liberated from the weed host disseminated by wind infects the fresh rice crop. Thus the weed hosts help to bridge the gap between two rice crops. Hence the pathogen can be able to line continuously in the vicinity on these hosts inspite of the non-cropping period intervening between two cropping periods. Intensive cultivation of particular crop repeatedly and constantly also provides perpetual inoculum. Powdery mildew and viral diseases of curcurbits are also best examples, where, number of cultivated crops serve as collateral hosts. The survival of the plant pathogens on collateral hosts/ alternative hostsoo which include the weed hosts also.

The collateral/weed hosts which are present in the field and in the hunds harbour the plant pathogens during cropping season. But the collateral/weed hosts present in the bunds harbour the plant pathogens during off-season. The pathogens can survive in active sporulating stage on wild collateral hosts and from there the primary inoculum may be disseminated by wind or insect to the main cultivated crops. Plant pathogenic bacteria may be able to survive in the parasitic form on annual and perennial weeds.

For example, the long term survival of Pseudomonas avenae in Florida is attributed to the association with a perennial grass, vaseygrass(Paspalum urvillei), through repeated infection of its vegetative growth as well as seed transmission. Bacterial leaf blight of maize may hae its origin in the infected vaseygrass distributed in the field.

Alternate hosts: The role of alternate hosts is not as important as of collateral hosts. However, when a pathogen has very wide host-range and is tolerant to wide range of weather conditions the alternate hosts become very important source of survival of the pathogen. These alternate hosts are very important for the completion of the life cycle of heteroecious rust pathogens. e.g. in temperate regions the alternate host Berberis vulgaris of Puccinia graminis tritici (black/stem rust pathogen on wheat), the barberry bush, grows side by side with the cultivated host, wheat. In such areas this wild host belonging to a different family is important for survival of the fungus. It helps in completion of heterogeneous infection chain of the rust fungus.

Self sown crops: Self sown plants, voluntary crops and early sown crops are reservoirs of many plant pathogens e.g., groundnut rust pathogen, Puccinia arachidis and ring mosaic of groundnut caused by tomato spotted wilt virus. Self sown rice plants harbour the pathogen as well as vector. e.g., rice tungro virus and its vector, Nephotettix virescens. Perennial crops also play a major role in the survival of the plant pathogens.

Pathogens which infect perennial plants can survive in them during low winter temperature and/or during the hot, dry weather of the summer. They survive in the lesions on perennial host plants which may be actively growing or are dormant. Disease like bunchy top of banana survives continuously in the suckers produced by the mother plants. In citrus canker the bacterium, Xanthomonas axonopodis pv. citri survive in the cankers formed on the leaves and twigs of the standing tree. They mostly survive in mild or vigorous active form on hosts, Citrus sp. Other examples, are Erwinia amylovora and Xanthomonas campestris pv. pruni.

Ratoon Crops: Sometimes ratoon crop also harbour the plant pathogens e.g., sugarcane mosaic.

Survival by latent infection: Latent infection refers to the conditions in which the plant pathogens may survive for a long time in plant tissue without development of visible symptoms. Eg. Pseudomonas syringae pv. syringae and X. axonopodis pv. citri can survive in apparently healthy bark tissues of their tree hosts. Xylella fastidiosa, the causal agent of Pierce's disease of grapevine and leaf scorch disease of various fruits and leaf ornamental trees infect diverse kinds of weeds without developing visible symptoms. Because these weeds are usually favourable habitats for the vector insects, latently infected weeds become an important source of the carrier insects.

Survival as residents: Plant pathogenic bacteria have the capacity to grow on the surface of host and non-host plants utilising the small amount of nutrients that are secreted on the plant surface.

Nematode and Fungi

Plant viruses like wheat mosaic, wheat spindle-streak virus, lettuce big vein, tobacco necrosis, tobacco rattle and tobacco ringspot viruses survive with nematodes or fungi found in the soil between crop seasons. Tobacco ringspot virus is associated with the nematode, Xiphinema americana.

The fungus, Polymyxa graminis (Barley yellow mosaic, oat yellow mosaic, wheat soil-borne mosaic, wheat spindle-streak mosaic) and Spongospora subterranea (potato mop top) carry the viruses internally and transmit them through their resting spore. Viruses are retained by nematode vectors for long times (stable). Xiphinema sp. retained viruses for a considerable length of time, while Longidorus spp. and Trichodorus spp. retained them for a much less period of one to two months only.

Insects

Many insects are carriers of inocula during the growing season and several important plant pathogens survive between growing seasons within insects. Some bacterial plant pathogens may survive within the insect body and over winter therein. The corn flea beetle, Chaetocnema pulicaria Melsh carries inside its body, the corn wilt pathogen, Xanthomonas stewartii and thus helps in its overwintering. The cucumber beetles, Diabrotica vitata Fabr. and D. duodecimpunctata Oliv., which chew the plant parts affected by Erwinia tracheiphila carry the pathogen inside their body, where it overwinters.

In the following seasons the insect passes on the bacterial pathogen to the host plant. The bacterial pathogen causing wilt of cucurbits is effectively transmitted by these insect vectors. It is reported that the bacterial pathogen and the insect vectors live in a symbiotic relationship, the insects helping the bacterium with protection from adverse weather conditions and the bacterium helping the insects with a supply of some digestive enzymes while it is inside the insect's body.

Plant viruses and phytoplasmas multiply within the vectors and can overwinter in those insects. Semi-persistent viruses are retained in the vectors for periods ranging from hours to days. Example, citrus tristeza virus is retained in the aphid, Toxoptera citricida. Persistent viruses retain the viruses from days to week. Most of the hopper-borne viruses multiply in their vectors. Viruses are retained through the moult and the vectors frequently remain viruliferous for life important vectors which retain the viruses are given below.

	Vector	Virus
Leafhopper	Circulifer tenellus	Beet curly top virus
Plant hopper	Cicadulina mbila	Maize streak virus
Green leaf hopper	Nephotettix cincticeps	Rice dwarf virus
Brown plant hopper	Nilaparvata lugens	Rice grass stunt virus
Hopper	Agallia constricta	Wound tumour virus

Transovarial transmission of the virus to the eggs of the vectors occurs, and the virus can multiply within a viruliferous hopper even if the insect is feeding on an immune host plant. Eggs carrying viruses may overwinter, and provide a source of virus to infect spring crops, even in the absence of diseased plants. Phytoplasmas attacking plants also multiply in the insects and remain infective throughout their life period. e.g. Rice dwarf virus (RDV) is transmitted through the eggs to about 60% of the progeny of the infective female leafhopper, Nephotettix cincticeps. RDV passes through the eggs to six succeeding generations.

Survival on Agricultural Materials

C. michiganensis subsp. michiganensis has been shown to

survive in air-dried conditions for 7 to 8 months on the surface of wooden stakes and boxes or wires or for 15 months in air-dried tissues of diseased tomato plants.

Survival on Surface Water

Erwinia carotovora subsp. carotovora is detected from water from drains, ditches, streams, rivers and lakes in mountainous upland and arable areas of Scotland and Colorado throughout the year..

Dispersal of Pathogens

Transport of spores or infectious bodies, acting as inoculum, from one host to another host at various distances resulting in the spread of disease, is called dissemination, dispersal or transmission of plant pathogens. It is very important for spread of plant diseases, for continuity of the life-cycle and evolution of the pathogen. The spores of some fungi are expelled forcibly from the sporophore or sporocarp by a squirting or puffing action that results in successive or simultaneous discharge of spores up to a centimetre or so above the sporophore.

The seeds of some parasitic plants are also expelled forcibly and may arch over distances of several metres. These are dispersed mechanically by various means. In bacterial diseases, the bacterial cells come out on the host surface as ooze or the tissues may be disintegrated so that the bacterial mass is exposed and then dispersed by various physical and biological agencies. Viral diseases which have no such organs are transmitted by insects, mites, phanerogamic parasites nematodes and human beings.

The knowledge of these methods of dispersal is essential for effective control of plant diseases because possibilities of preventing dispersal and thereby breaking the infection chain exist. The dispersal of infectious plant pathogens occurs through two ways,

1. Autonomous or direct or active dispersal
2. Indirect or passive dispersal

Autonomous Dispersal

It is also known as active or direct dispersal. In this method the dispersal of plant pathogens (fungi, bacteria, and viruses) takes place through soil and seed or planting materials during normal agronomic operations.

Soil as means of autonomous dispersal: Soil-borne facultative saprophytes or facultative parasites may survive through soil. The dispersal may be by movement of the pathogen in the soil or by its growth in soil or by movement of the soil containing the pathogen. The former is known as dispersal in soil while the latter is called dispersal by soil.

a. *Dispersal in soil* : The following are the three stages of dispersal in soil:
   i. Contamination of soil
   ii. Growth and spread of the pathogen in soil
   iii. Persistence of the pathogen
b. *Dispersal by the soil*: The pathogen enters the soil, grow and spread in the soil. During the cultural operations in the field, soil is moved from one place to the nearby place through the agricultural implements and irrigation, worker's feet. Propagules of fungi or the dormant structures of fungi and the plant debris containing the fungal and bacterial pathogens thus spread throughout the field. This type of dispersal is highly erratic. The

most important methods of dispersal of pathogen by the soil are transfer of soil from one place to another along with plant parts or propagating materials. e.g., transfer of papaya seedlings from a nursery infested with Pythium aphanidermatum (the cause of stem or foot rot of papaya) can introduce the pathogen in new pits for transplanting the seedlings.

Similarly grafts of fruit trees transported with soil around their roots can transmit pathogens present in the nursery to the orchards. By this method, pathogens are not only spread from field to the field but also from district to district, State to State and often from country to country.

Seed and seed materials as the source of autonomous dispersal: The seeds serve as medium for autonomous dispersal of pathogens. Since most of the cultivated crops are raised from seed the transmission of diseases and transport of pathogens by seeds has much importance. The dormant structures of the pathogen (e.g., seeds of Cuscuta, sclerotia of ergot fungus, smut sori, etc.,) are found mixed with seed lots and they are dispersed as seed contaminant.

The bacterial cells or spores of fungi present on the seed coat (such as in smuts of barley, sorghum, etc.,) are transported to long distances. Dormant mycelium of many fungi present in the seed is transmitted to long distances. There are three types of dispersal by seed viz.:

Contamination of the seed: Seed -borne pathogens move in seed lot as separate contaminants without being in intimate contact with the viable crop seeds. The seeds of the pathogen or parasite and the host are getting mixed during harvest of the crop. In many cases, the identity of the seeds of the two entities (host and the parasite) is difficult to separate. e.g., smut of pearlmillet (Tolyposporium penicillariae), ergot of rye and pearlmillet (Claviceps purpurea and C. microcephala respectively).

Smut soil and ergots mix easily with the seed lots during harvest or threshing, In many smuts such as Karnal bunt of wheat (Neovossia indica) and bunt of rice (Neovossia horrida) the infected kernels containing smut sori are mixed with the seed. In leaf smut of rice (Entyloma oryzae) leaf pieces containing smut sori are mixed with the seed.

Externally seed-borne: Close contact between structure of the pathogen and seeds is established in diseases like covered smut and loose smut of barley, short smut of sorghum, stinking smut of wheat and bacterial blight of cotton where the pathogen gets lodged in the form of dormant spores or bacteria on the seed coat during growth of the crop or at the time of harvest and threshing. In many pathogens the externally seed-borne structures such as smut spores can persist for many years due to their inherent capacity for long survival. The spores of Tilletia caries (stinking smut of wheat) remain viable even after 18 years and those of Ustilago avenae (oat smut) for 13 years.

Internally seed-borne: The pathogen may penetrate into the ovary and cause infection of the embryo while it is developing. They become internally seed-borne. Internally seed borne pathogens like Ustilago nuda tritici are viable for many years.

Seeds of cultivated crops are distributed mainly by man. Sometimes animals and birds also help in distribution of crop seeds. Man and animals are the main agencies of dispersal of pathogen through seed. The pathogens thus mixed with the seed or on the seed are transmitted.

Passive Dispersal

Animate Agents

Insects: Insects carry plant pathogens either externally or internally. Gaiiman used the terms epizoic and endozoic respectively for these two types of transmission. The external transmission of plant pathogens is of special interest in those fungi, which produce their conidia, oidia and spermatia in sweet or honey secretions having attractive odours. Some of the well known diseases of this type are the ergot, the Sclerotinia brown rot of pear and apple, the honey dew stage in the 'sugary disease' of sorghum and pearlmillet in parts of India and the pycnial nectar in the cluster cup stage of rusts.

The spermatial oozing at the mouth of spermagonia in the Ascomycetes attract various types of insects, flies, pollinating bees and wasps which play a dual role viz., pollination and transmission of pathogens. The fire blight organism (Erwinia amylovora) pathogens and citrus canker bacterium, (Xanthomonas axonopodis pv. citri) are also carried in this manner, the former by flies (bees) and ants and the latter by the leaf miner. The black leg of potato caused by Erwinia carotovora is disseminated by maggots, wilt of corn caused by X.stewartii, gummosis of sugarcane caused by X.vasculorum are the other examples for bacterial diseases transmitted by insects.

Ingenious transmission of pathogens, of an internal nature (endozoic) is provided by the Dutch elm disease (Ceratostomella ulmi) and the olive canker (Bacillus savastanoi). The former is transmitted by the elm bark beetles and the latter by the olive fly (Olea europaea).

These insects, unlike the epizoic group, appear to have a close biologic relationship with the pathogens, as they have not been reared without the contaminating pathogens. Few important plant pathogenic bacteria are spread by insects.

The cucumber wilt bacterium, Erwinia tracheiphila is spread by the striped cucumber beetles (Acalymma vitata) and the spotted cucumber beetle (Diabrotica undecimpunctata). When the beetles are feeding on the diseased plant, the bacterium contaminates the mouth parts and passes into the gut of the insect. During the winter season, the bacterium overwinters inside the beetle. Thus the beetle helps the bacteria in two ways, i.e in their transmission and survival. More than 80 per cent of the viral and phytoplasmal diseases are spread by different types of insects. The insect which act as specific carriers in disseminating the diseases are called insect vector.

Both aphids (Aphidae) and leaf hoppers (Cicadellidae or Jassidae) in the order Homoptera contain largest number and the most important insect vectors of plant viruses. Certain species of mealy bugs and scale insects (Coccoidae), whiteflies (Aleurodidae) and treehoppers (Membracidae) in the same order (Homoptera) also transmit virus diseases. Insect vectors of plant viruses are few in true bugs (Hemiptera), thrips (Thysanoptera), beetles (Coleoptera) and grasshoppers (Orthoptera). Aphids, leafhoppers and other groups of Homoptera and true bugs have piercing and sucking mouth parts.

Thrips have rasping and sucking mouth parts. All other groups of insect vectors have chewing mouth parts and they transmit only very few viruses. Aphids: Aphids are the most important insect vectors of plant viruses and transmit the great majority of all stylet—borne viruses. As a rule several aphid species can transmit the same stylet—borne virus and the same aphid species can transmit several viruses, but in many cases the vector-virus relationship is quite specific.

Aphids generally acquire the stylet-borne virus after feeding on a diseased plant for only a few seconds (30 seconds or less) and can transmit the virus after transfer to and feeding on a healthy plant for a similarly short time of a few seconds. The length of time aphids remain viruliferous after acquisition of a stylet-borne virus varies from a few minutes to several hours, after which they can no longer transmit the virus.

In few cases of aphid transmission of circulative viruses, aphids cannot transmit the virus immediately but must wait several hours after the acquisition feeding, but once they start to transmit the virus, they continue to do so for many days following the removal of the insects from the virus source. In aphid transmitting stylet-borne viruses, the virus seems to be borne on the tips of the stylets, it is easily lost through the scouring that occurs during probing of host cells, and it does not persist through the moult or egg.

Leafhoppers: Leafhoppers are phloem feeders and acquire the virus from the phloem region. All leafhopper, transmitted viruses are circulatory. Several of these viruses multiply in the vector (propagative) and some persist through the moult and are transmitted through the egg stage of the vector. Most leafhopper vectors require a feeding period of one to several days before they become viruliferous, but once, they have acquired the virus they may remain viruliferous for the rest of their lives. Usually there is an incubation period of 1 to 2 weeks between the time a leafhopper acquires a virus and the time it can transmit it for the first time.

Fungi: Some soil—borne fungal plant pathogens transmit plant viruses. Olpidium brassicae, Ploymyxa graminis, P. betae and Spongospora subterranea are the fungi involved in transmission of virus disease. The viruses are apparently borne in or on the resting spores and the zoospores, which upon infection of new host plants introduce the virus and cause symptoms characteristic of the virus they transmit. All these fungi are pathogens of the host which carry of viruses.

The zoospores of the fungi are released from the host and the zoospores carry the virus and transmit it to the susceptible hosts during their infection process. In some cases plant viruses are carried on the outside of the fungi. Examples are tobacco necrosis virus and cucumber mosaic virus. The viruses like lettuce big vein virus are found inside the zoospores. They persist for years in viable resting sporangia. The types of transmission by fungi can be considered as non-persistent and persistent transmission.

Nematodes: Nematodes are soil borne organisms. Some of the nematodes act as agents for dissemination of pathogenic fungi, bacteria and viruses. For example, the bacterium Corynebacterium tritici which causes yellow ear rot of wheat is disseminated by ear cockle nematode. Similarly, some pathogenic fungi such as, Phytophthora, Fusarium, Rhizoctonia, etc., are carried on the body of nematodes. Nematodes help these pathogenic fungi to enter into the host through punctures for their own entry and enter into hosts along with the nematodes.

Plant nematodes play a vital role as vector in transmitting certain virus diseases. Nematode vectors transmit viruses by feeding on roots of infected plants and then moving on roots of healthy plants. Larvae as well as adult nematodes can acquire and transmit viruses, but the virus is not carried through the larval molts or through the eggs. After moulting, the larvae or the resulting adults must feed on a virus source before they can transmit again. Both the polyhedral and tubular type of viruses are transmitted by Xiphinema, Longidorus and Trichodorus.

Human beings: Man is the most important factor responsible for short distance and long distance dispersal of plant pathogens. He helps in dissemination unknowingly by his usual agricultural practices. Human being's role in dissemination of plant pathogens is more direct than indirect. The ways and means by which human beings help in dispersal are as follows.

a) *Seed trade*: Seed trade is one of the different means of dispersal of plant pathogens in which man plays an important role. The import and export of contaminated seeds without proper precautions lead to movement of pathogens from one country to another or from one continent to another. Through this way pathogens of soybean and sugarbeet hither to not prevalent in India got introduced.

   Human agencies of individual, official and unofficial have transported new plants and plant products, the seed, the tubers, the propagating stock and fruits, which carried the plant pathogens, many times in a latent condition and which ultimately lead to the outbreaks of new diseases in places, hither to free from them. The diseases which are amenable to such transmission are mainly those that are carried in or on the propagative parts and seed such as late blight of potato, the downy mildew of grapevine, citrus canker, chestnut blight, Dutch elm disease, Fusarium wilt of banana, Katte disease of cardamom and bunchy top of banana.

b) *Planting diseased seed materials (vegetatively propagated materials)*: Planting diseased bulbs, bulbils, corns, tubers, rhizomes, cutting etc. of vegetatively propagated plants such as potato, sweet potato, cassava, sugarcane, banana, many ornamentals and fruit trees etc. help in dispersal of pathogens from field to field, orchard to orchard, locality to locality or from one country to another.

c) *By adopting farming practices*: Human beings (men and women) engaged in preparatory cultivation, planting, irrigation, weeding, pruning etc. help in dispersal of plant pathogens. The fungal spores (oospores, chlamydospores), dormant structures like sclerotia are carried by worker's clothing, shoes, hand etc. from field to field. Men or women engaged in intercultivation in tobacco field spread the dreaded tobacco.

d) *Through clothing*: Palm workers engaged in cleaning coconut trees spread bud rot disease.

e) *By use of contaminated implements*: Pathogens are transferred from one area to another through implements used in various cultural operations (weeding, hoeing thinning etc.) in the field. e.g. root rot of pulses and cotton (Macrophomina phaseolina, bacterial angular leaf spot of cucumber (Pseudomonas lachrymans) and bacterial canker of tomato (Corynebacterium michiganensis). Cutting knives and pruning knives help in dissemination from one plant to another e.g., Bunchy top of banana.

f) *By use of diseased grafting and budding materials*: Grafting and budding between healthy and diseased plants is the most effective method of distribution of pathogens of horticultural crops (fruit trees, ornamentals etc.) e.g., Careless selection of stocks and scions in propagation of citrus trees.

Farm and wild animals: Farm animals (cattles) while feeding on diseased fodder ingest the viable fungal propagules (spores or oospores or sclerotia) into their digestive system. Animals which feed on downy mildew affected pearlmillet or sorghum take the oospores along with the fodder. Oospores pass out as such in the dung. This dung when used as manure spread in the field and act as source of inoculum.

Smut fungi like grain smut of sorghum, loose smut and head smut of sorghum are carried from field to field through the alimentary canals of farm animals. Soil inhabiting fungi especially sclerotia adhere to the hoofs and legs of animals and get transported to other places. Animals passing through the tobacco fields help in transmission of TMV.

Birds: In general, transmission by birds is of minor importance. But this method is important in dissemination of seeds of flowering parasites and certain fungi. Many migratory birds, such as mistle thrush (Turdus viscivorus) in the temperate region and the crows (Crovos brachyrhynchos) in the tropics, take active part in the transmission of giant mistletoe (Dendrophthoe spp.) either through external contamination of their beaks and feathers or internally through the alimentary canals.

These birds feeding on the fleshy, sticky and gelatinous berries of giant mistletoe deposit the seeds on the other trees with the excreta. Stem segments of dodder (Cuscuta spp.) are carried by birds for building their nests. Thus the phanerogamic parasites are getting transported to new locations. Spores of chestnut blight fungus, Endothea parasitica are disseminated by not less than 18 species of birds. Internal transmission of this pathogen is carried out by the birds which visit such diseased plants and get contaminated by the spores. Birds are also known to carry the spores of fungi on their body.

Phanerogamic parasites: Plant viruses are transmitted from one plant to another through the bridge formed between the two plants by the twining stems of the parasitic plant dodder (Cuscuta spp). Dodder is yellow vine without green leaves. In this way viruses are transmitted between plants belonging to families widely separated taxonomically. The virus is transmitted in the food stream of the dodder plant, being acquired from the vascular bundles of the infected plant by the haustoria of dodder.

After translocation through the dodder phloem the virus is introduced in the next plant by the new dodder haustoria produced in contact with the vascular bundles of the inoculated plant. Cuscuta californica, C. campestris, C. subinclusa are usually employed for dodder transmission of viruses and phytoplasmas. C. europaea, C. epilinum and C. lupuliformis are also employed in transmission of viruses.

Inanimate agents

Wind: The wind dispersal of plant pathogens is known as anemochory. It is one of the most common methods of the dispersal of plant pathogens. It is the most dangerous and potent mode of travel for plant pathogenic fungi. It acts as potent carrier of propagules of fungi, bacteria and viruses. Usually the fungal pathogens are light in weight and are well adapted to wind dispersal. Some pathogenic bacteria are carried along with the infected material to short distances by wind. Damping-off pathogen (Pythium spp.), wart disease pathogen of potato (Synchytrium endobioticum), root rot pathogens (Sclerotium and

Rhizoctonia) and seeds of phanerogamic parasites witchweed (Striga) are efficiently carried by wind.

Viruses and phytoplasmas are not directly transmitted by wind, but the insect and mite vectors that carry the viruses move to different directions and distances depending upon the direction and speed of air. The adaptations for wind dispersal in fungal pathogens include, production of numerous spores and conidia, discharge of spores with sufficient force, production of very small and light spores so that they can move to long distances. The duration and periodicity of sporulation and discharge are also important factors for wind dispersal.

Some fungal pathogens causing powdery mildews, downy mildews, rusts, smuts, sooty moulds, leaf spots, blast, apple scab etc., produce large number of very light spores and conidia on the surface of the host. Uredial stages of the rust fungi travel long distances through air currents and are thus responsible for destructive epidemics over wide areas. Wind transmission involves the upward air currents, velocity and the downward movements of wind. All are equally responsible for the spread of infection and ultimate outbreak of diseases and have been of special significance in the rust, smut and blast fungi.

Urediospores of rust fungi have been carried to long distances, both cross-wise and upwards. The transmission of aecial spores of Puccinia graminis tritici from several groups of barbery bushes to the wheat crop showed that these spores traveled successfully over a radius of 3 kms round about these bushes. The blister rust fungus, Cronartium ribicola, is known to travel to a distance of 500 metres or 3,750 m. inside a plantation that the range is probably more in the open. Similar observations have been made in respect of dissemination of chlamydospores of the smut fungi.

In long distance dissemination with intervening stages of infection, the retention of viability of spores is an important factor that determines the extent and severity of epidemics, over wide areas. The outbreaks of cereal rusts and blast of rice are examples of such dissemination. Spores differ widely in their ability to survive long distance travel through air. Urediospores of rusts, chlamydospores of smut fungi and conidia of Alternaria, Helminthosporium, Pyricularia and others are well adapted for long distance travel in a viable condition and are known to play a vital role in epidemiology.

The conidia of downy mildews, powdery mildews and the aeciospores and basidiospores of the rust fungi are unable to withstand such long distance dissemination when they are exposed to desiccation and direct sunshine and thus are only capable of producing local epiphytotics of limited magnitude. The bacteria causing fire blight of apple and pear (Erwinia amylovora) produce fine strands of dried bacterial exudate containing bacteria and these strands may be broken off and they are transmitted by wind. Bacteria and nematodes present in the soil may be shown away along with soil particles.

Wind also helps in the dissemination of bacteria, fungal spores and nematodes by blowing away rain splash droplets containing these pathogens. Wind carries insects and mites that may contain or are smeared with viruses, bacteria or fungal spores to short or long distances. Wind also causes adjacent plants or plant parts to rub against each other. The wound created in this manner help the spread by contact of bacteria (citrus canker), fungi, some viruses (Tobacco mosaic virus) and viroids and possibly of some nematodes.

Water: Transmission of plant pathogens by water is not as significant as wind transmission. Although water is less important than air in long-distance transport of pathogens, water dissemination of pathogens is more efficient, in that the pathogens land on an already wet surface and can move or germinate immediately. In case of some diseases the surface flow of water after heavy showers of rains or irrigation water from canals and wells carries the pathogens to short distances. Soil-inhabiting fungi like, Fusarium, Ganoderma, Macrophomina Phytophthora, Plasmodiophora, Pythium, Rhizoctonia, Sclerotium, Sclerotinia, Sporisorium, Ustilago, Verticillium etc., in the form of mycelial fragments, spores or sclerotia, soil-bone bacteria and nematodes carrying viruses are transmitted through the above process. They are transmitted through rain or irrigation water that moves on the surface or through the soil.

All bacteria and the spores of many fungi are exuded in a sticky liquid and depend for their dissemination on rain or (overhead) irrigation water which either washes them downward or splashes them in all directions. Raindrops or drops from overhead irrigation pickup the fungal spores (uredospores of Homilies, Puccinia and Uromyces and bacteria (bacterial blight pathogen of rice, Xanthomonas oryzae pv. oryzae; bacterial leaf streak pathogen, X.oryzae pv. translucens; citrus canker pathogen, X.axonopodis pv. citri; tomato bacterial blight pathogen, Corynebacterium michiganese and cotton bacterial blight pathogen, X.axonopodis pv. malvacearum present in the air and wash them downward where some of them may land on susceptible plants.

Epidemiology

Edpidemiology or epiphytology is the study of the

outbreak of disease, its course, intensity, cause and effects and the various factors governing it. Disease may affect isolated individuals within a crop and is called sporadic. On the other hand, if the disease assumes destructive nature over widespread areas, involving sudden outbreaks it is known as epidemic or epiphytotic. When a host and the pathogen reach a state of biologic equilibrium and assumes apparently harmless character, it is termed endemic.

An epidemic may cause widespread and mass destruction of crop in a short time or may persist for long periods depending upon the three following factors responsible for the disease:

1. Host
2. Pathogen and
3. Environment

A course of epidemic in nature differs with the nature of the host, the pathogen and the environment. Epidemics caused as a result of introduction of new pathogens in the locality hither to free from them, appear in two phases viz., destructive phase and innocent phase (due to biologic equilibrium reached between new comer pathogen and the original inhabitant). The well known epidemics of late blight of potato in Europe and blast disease of rice in South East Asia, powdery mildew and downy mildew of grapevine in Europe, leaf rust of coffee in Sri Lanka and anthracnose of grapevine in India are examples of this category. In the above diseases the pathogens after taking heavy toll of the crops have settled down.

A disease is sometimes sporadic and assumes epidemic proportions under special circumstances. The essential conditions for an epiphytotic or the factors governing epidemics can be grouped under the three heads.

1. Nature of host
2. Nature of the pathogen and
3. Environment

An epidemic can only result from the cumulative effects of all the three factors mentioned above, acting simultaneously. Few pathogens are capable of assuming epiphytotic conditions while others are sporadic. The former group consists of late blight of potato, blast of rice, downy mildew diseases and rust diseases.

Forecasting of Diseases

Forecasting of plant diseases means predicting for the occurrence of plant disease in a specified area ahead of time, so that suitable control measures can be undertaken in advance to avoid losses. Disease forecasts are predictions of probable outbreaks or increase in intensity of disease. It involves well organised team work and expenditure of time, energy and money. It is used as an aid to the timely application of chemicals. Among the first spray warning services to be established for growers, were the grapevine downy mildew forecasting schemes in France, Germany and Italy in the 1920s. Disease forecasting methods are available for the following plant diseases.

Forecasting diseases is a part of applied epidemiology. Hence, knowledge of epidemiology (development of disease under the influence of factors associated with the host, pathogen) is necessary for accurate forecasting. The factors of epidemic and its components should be known in advance before forecasting is done. The informations required for forecasting are:

1. Host Factors
    a. Prevalence of susceptible varieties in the given locality
    b. Response of host at different stages of the growth to the activity of pathogen e.g. Some diseases are found during seedling stages while others attack grown up plants and
    c. Density and distribution of the host in a given locality. Dense populations of susceptible variety invite quick spread of an epidemic. Growing susceptible varieties in scattered locations and that too in a limited area are less prone to epiphytotic.
2. Pathogen factors
    a. Amount of primary (initial) inoculum in the air, soil or planting material
    b. Dispersal of inoculum
    c. Spore germination
    d. Infection
    e. Incubation period
    f. Sporulation on the infected host
    g. Re-dispersal / Dissemination of spores
    h. Perennating stages
    i. Inoculum potential and density in the seed, soil and air
3. Environmental factors
    a. Temperature

b. Humidity

c. Light intensity

d. Wind velocity

There are five main requirements which must be satisfied before a useful and successful disease forecast is made.

1. The disease must cause economically significant damage in terms of yield loss or quality: Damage assessment is essential to develop strategy for controlling a disease. e.g., Annual estimation of yield loss caused by barley powdery mildew (Erysiphe polygoni) in England and Wales had ranged from 6 to 13 %. Potato late blight can cause a yield loss of 28% if the disease reaches the 75% stage by mid-August. Diseases like apple scab and potato common scab reduces the quality of the produce lower the value of the harvested crop and cause considerable financial loss to the growers.

2. Control measures must be available at an economically acceptable cost.

3. The disease must vary each season in the timing of the first infections and its subsequent rate of progress. If it does not, there is no need for forecasting.

4. The criteria or model used in making a prediction must be based on sound investigational work carried out in the laboratory and in the field and tested over a number of years to establish its accuracy and applicability in all the locations where its use is envisaged.

5. Growers must have sufficient man power and equipments to apply control measures when disease warning is given. Long-term warnings or

predictions are more useful than short-term warning or predictions.

Forecasting Methods

Disease forecasting requires field observations on the pathogen characters, collection of weather data, variety of the crop and certain investigations and their correlations. Usually the following methods are employed in disease forecasting.

1. *Forecasting based on primary inoculum*: Presence of primary inoculum, its density and viability are determined in the air, soil or planting material. Occurrence of viable spores or propagules in the air can be assessed by using different air trapping devices (spore traps). In the case of soil-borne diseases the primary inoculum in the soil can be determined by monoculture method. Presence of loose smut of wheat, ergot of pearlmillet and viral diseases of potato can be detected in the seed lots at random by different seed testing methods. Seed testing methods can be used to determine potential disease incidence and enable decision to be made on the need for chemical seed treatment. The extent of many virus diseases is dependent on the severity of the preceding winter which affects the size of vector population in the growing season. e.g., Sugarbeet yellows virus.

2. *Forecasting based on weather conditions*: Weather conditions viz., temperature, relative humidity, rainfall, light, wind velocity etc., during the crop season and during the inter crop season are measured. Weather conditions above the crop and at the soil surface are also recorded.

3. *Forecasting based on correlative information*: Weather data of several years are collected and correlated with the intensity of the diseases. The data are compared and then the forecasting of the disease is done. Forecasting criteria developed from comparisons of disease observation with standard meteorological data have been provided for diseases like Septoria leaf blotch of wheat, fire blight of apple and barley powdery mildew.
4. *Use of computer for disease forecasting*: In some advanced countries forecasting of disease is made by the use of computers. This system gives the results quickly. One such computer based programmes in the USA is known as 'Blitecast' for potato late blight.

Spore trapping

Techniques of acquisition of biological data for consecutive forecasting models are important. Spore traps have been widely used in to complete disease with weather conditions. Spore trapping is useful for understanding epidemiology of a disease and behaviour of the pathogens. This helps in developing models on dispersal of pathogens or on epidemiology of the disease and to formulate methods of management. Methodology of spore trapping depends on the following objectives of the worker.

1. Biology of the pathogen
3. For infection forecasting
2. Spore dispersal gradients
4. Management of the disease

In epidemics of air-borne plant diseases the number of spores of the pathogen landing on the plant which depends on the number of spores in the atmosphere above the crop is an important factor for the quantitative sampling of the atmosphere (number of spores per unit volume of air). For trapping and estimating these studies different types of traps are used. The following spore traps are usually employed in trapping of fungal spores.

1. *Cylindrical rods or microscopic glass slides*: It helps to gather data on the spore arrival in a locality. In this, the surface of microscopic slide is smeared with grease and made sticky. In the method, quantitative estimation is not possible as number of spores collected is very low.
2. *Hirst's volumetric spore trap*: In this instrument, air is sucked into at a controlled rate and impinged on to a glass slide moved by a clockwork mechanism past the orifice. It gives continuous count of spores in 24 h. The number of spores per unit volume of air at any given time can thus be calculated.
3. *Rotorod sampler or rotorod spore trap*: It comprises of a 'U' shaped rod attached at its mid point to the shaft of a small battery operated electric motor. In this equipment the surface of the rod is covered with a vaseline strip of transparent cellophanes to catch spores which can be taken off and mounted on a glass slide. From the area of the strip and the speed of rotation, the volume of air samples can be calculated.
4. *Anderson cascade spore sampler*: It is a device where Petri plates with nutrient agar are used to collect the spores.
5. *Bourdillon slit sampler*: Air is sucked in a chamber by vacuum pump which strikes the rotating Petri dish containing agar medium. The agar medium retains

the spores sucked in the air. Concentration of viable spores is calculated after counting germinated spores in the medium.

6. *Burkard's 7 day volumetric spore trap*: This device records spores in the air drawn by a pump on 7 days basis on a cellophane strip wrapped on a drum rotating inside a chamber.
7. *Jet spore trap*: In the above sampling methods, the viability of the spores cannot be determined. To overcome this, living plants have been used as spore traps. A jet spore trap in which spores are impacted in an air jet into a column of still air, through which they fall, to settle on leaf segments exposed at the base of the chamber. In this trap, suitable cultivars of host plants can be employed to determine number of viable spores.

5

Agro-Biodiversity and Seed Security

Agricultural biodiversity or Agro-biodiversity is a sub-set of general biodiversity including all cultivated varities. Agricultural biodiversity contributes to food security and livelihood security and underpins the development of all food production. It is the first link in the food chain, developed and safeguarded by farmers, livestock breeders, forest workers, fishermen and indigenous peoples throughout the world.

The agro-biodiversity could be viewed from two contrasting angles, namely, (i) the traditional agro-eco-systems, and (ii) the intensive farming systems. Whereas the diversity among the cultivated plant species is maintained, more or less in equilibrium in the former case, it has to be guided externally in case of the latter. It is estimated that 20 percent of world's current food production is met from traditional multi-cropping systems which also help in maintaining ecological equilibrium and sustaining crop genepool admixtures, transgression and other micro-evolutionary processes.

The intensive cropping systems or the modern agro-eco-systems, on the other side, have evolved more recently, basically to enhance production vis a vis farm economic returns. A remarkable success has been shown in India harnessing the natural resources coupled with development and utilisation of improved technology. It included a 275 per cent increase in the foodgrains production in the last fifty years, which resulted in a buffer stock of 36 million tonnes in 1994-95 and a record harvest of 199 million tonnes in 1996-97 as against a total production of only 51 million tonnes in 1950-51. This achievement was made in spite of a growing population pressure at 2.1 percent per annum.

The estimates showed that a meagre 8 per cent increase in food production has been achieved due to increase in area under cultivation whereas the rest 92 percent increase resulted from higher yield per hectare which has apparently been contributed by improved varieties, better agronomic packages of cultivation practices and better disease and pest control measures. The selection of improved types in all these cases could be made possible due to the availability of diversity along with expertise to select the right 'types'.

Estimates also showed that, on an average, more than half of the cultivated areas under major foodgrain crops, viz., wheat, rice, sorghum, maize and pulses were under high yielding varieties. Production in these systems is more dependent on the responsiveness of improved cultivars to various inputs, farm mechanisation, etc. and the various crops and varieties have coexisted albeit based on relatively limited research results.

Several threats to agro-biodiversity have recently emerged in the advanced agricultural systems, particularly in many technologically advanced countries which posed several questions. Some of the relevant issues that emerged include:

i) safe conservation of landraces/farmers consciously selected heritage seeds and their management under a sustained system providing for further opportunities for agro-evolutionary processes; and

ii) an efficient, judicious and equitable use of this diversity to attain progressive increases in yielding ability and desired adaptability of improved varieties.

In this entire chain of events, seed of quality holds the key to ensure other agro-production inputs to be far more productive and cost effective. One can not think of an effective food and nutritional security unless the seed, being the basic input to crop oriented production, is assured.

Sustaining Agro-biodiversity

Nations were required to evolve a sui generis system for varietal protection, go in for patenting or adopt a combination of both, but not in isolation from the broader goals set out in agenda 21 of the CBD, particularly chapter 8. Such a sui generis system, in absolute sense, provides an institutional mechanism and legislative protection to manage and enrich inter alia the agro-biodiversity, within the national territory. It would give due opportunities to on-farm conservation for sustainable use besides protecting and safeguarding the interests of those who conserve and maintain the vital components.

Issues that the nations must consider, as per the requirements of CBD, include to plan and implement in an effective legal and/or regulatory frame work which would ensure no harmful effects even on territory beyond national borders upon consequent use of various components of agro-biodiversity; the regulation of exchange of germplasm is done in conformity with the charter of the UN and the international law, and that an intellectual property protection is provided to plants or plant parts in other member countries, etc.

Nations may be required to maintain two contrasting agro-ecosystems. First, the traditional agro-ecosystems encompassing and protecting biological diversity therein along with its biotic, abiotic and environmental components providing/maintaining due opportunities to sustainability and the micro-evolutionary processes. The cost of maintaining such traditional agro-ecosystems, through imparting awareness and providing suitable compensations to the local folk who would practise it, is bound to be high but looking into the indirect impacts of such conservative mechanisms particularly on the climatic equilibrium, the CBD appealed to Global Environmental Fund (GEF), in one of the decisions made in the third meeting of the Conference of Parties (COP-III), to make liberal financial grants for the above purposes pending settlement of the issue of establishing a Global Biodiversity Fund, which is being strongly advocated by several partners and fora.

Lessons could also be learnt from the more recent indigenous knowledge and social framework of traditional monospecific cropping systems so as to help maintain the ecological equilibrium in advance, highly productive, intensive agro-ecosystems. Funding provisions may specifically be made for conducting research on these aspects.

A concern has been shown that the traditional agro-ecosystems should be conserved along with diverse evolutionary forms of cultivated plant species as well as farmers' conscious selections, particularly in the areas of origin and diversity of crop plant species and their relatives. This, obviously calls for concerted efforts towards developing a better understanding in terms of the following:

i) classification, indexing and inventorisation of traditional agricultural systems, still in vogue, with respect to their ecosystems, crop combinations, seasons and/or systems of cultivation;

ii) potential core area(s) under each, respective system or sub-system; and

iii) the custodian community.

At the same time, guiding and maintaining diversity in the advance agro-ecosystems through research, extension and legislation is considered essential to tackle several problem situations, such as, biotic virulences, abiotic stresses and adverse climates. Rapid appraisal systems are, therefore, essential so as to maintain and guide requisite diversity in these agricultural systems to enable fast detection and appropriate replacement of the varietal components, such as those succumbed to diseases/pests/abiotic stresses or those which have been surpassed by more productive types in research.

The creation and maintenance of agro-biodiversity could be ensured through multi-crop, multi-varietal mosaic cropping which shall mimic the co-existence of cultivated plant species and the within-species diversity in the traditional agro-ecosystems. Also, the maintenance of the latter through awareness generation and providing suitable compensation to the local people who practise it, would be essential for an effective management of agro-biodiversity.

In situ on-farm conservation

There is a demanding need to enumerate agreed scientific principles and practices for 'on-farm' genetic resource conservation. It includes usable classification and nomenclature systems for landraces, as well as methodology for their characterisation. Due to lack of available information of this type, it is imperative that studies should be carried out on these aspects side-by-side while implementing 'on-farm' conservation programmes. A better understanding of the temporal and spatial movement of genes and varieties into and out of cropping systems and the causes of loss or recession of some varieties or traits over others should be highly useful in further setting the pace and direction of crop evolution.

Adequate studies of some of the traditional practices and customs have not been carried out in most of the agro-biodiversity rich areas. Systematic surveys and collection of data on the traditional knowledge base, including methods to maintain healthy populations and seed, require an inquisitive follow-up and indexing. The 'on-farm' conservation is generally associated with traditional agro-ecosystems and small scale farming although the farmers' participatory approach has been more advocated to help select suitable high yielding varieties (HYVs) for the advance cultivation systems.

The large-scale commercial farms also have options to promote agro-biological diversity through crop rotations, intercropping, cover crops, green manures, paddy-cum-fish culture, apiculture, etc. The seed requirements for such situations would depend upon the adaptability, preference and acceptability of component HYVs for each sub-situation. Such practices need to be standardised as per diverse agro-ecological zones, so that sustainability of seed component is ensured under a dynamic intensive cultivation system.

The more likely negative impact on sustainability of seed under the intensive cropping system may occur due to successively superior varieties being evolved at frequent intervals, the imminent displacement of modified agro-ecosystems which were created by cultivating such superior HYVs, the evolution of virulences or other environmental factors, such as, off-site pollution or accumulation of certain harmful gases like methane or ozone, etc. which need to be addressed by well planned, self-generating mechanisms.

Domestication of individual crop plant species is unique to agro-ecological sub-regions and localities and corresponds to cultural and ethnic diversity, including food habits, habitats, climates and available crop plant species. There is a purposeful need to maintain their representative seed 'on farm' because re-building new varieties for similar situations with same amount of adaptability and phenotypic plasticity would require strenuous planning, breeding efforts and resources, including funds and time.

Capacity building for 'on-farm' conservation of traditional 'types' requires formulation of suitable policies, sectoral and cross-sectoral plans and programmes for implementation and also interim financial mechanisms. Recognition of the associated local knowledge, innovations and informal practices would be needed to promote traditional farming systems, whereas establishing community seed banks in co-operative or private sectors, developing and maintaining cultivar/ landrace registration system, etc., is required for smooth replenishment.

The recognition of 'Farmers' Rights' is important along with formulation of implementable laws and regulations so as to ensure availability of the much needed 'seed' with due compensation or assurance of benefit sharing to its holders. Country level implementation of Registration of Plant Germplasm has been implemented in India by the ICAR and the planning/action related to other aspects is also initiated.

Compiling case studies for conservative, low external output based production systems through incentives for developing environmentally sound technologies for conservation, replenishment and sustainable use of biological diversity are issues which lack direction and, therefore, need sincere input from all concerned under the sui generis situations.

Geo-sectoral approach

Contiguity of adjoining areas across the geo-political barriers, representing similar genetic diversity for particular species genepool, can be attributed as a basis to conserve, utilise and share the benefits of the in situ—on-farm conservation. Safeguarding the regional 'gene' interests could be ensured through sub-regional consortia and regional networking. The increasing role of Global Forum and the regional fora in strengthening relationships among the national agricultural research systems (NARS) and between various NARS and the Commonwealth Group of International Agricultural Research Institutions (CGIAR), could be viewed as new initiatives in desired direction which need to be pursued.

Technological and technical needs can be gradually fulfilled for innovating, standardising and researching into the rapid appraisal systems for agro-biodiversity management using and strengthening the information network(s), as well as database system(s). Due to the recent global interest for germplasm exchange and concern for protection of Plant Breeders'/Farmers' Rights there is urgency that the countries develop a regional consensus by way of the following:

i) by entering into agreement for undertaking joint explorations in the region for collecting genepools of crops for which the region is centre of origin;

ii) by adhering to effective plant quarantine so as to maintain an effective mechanism for exchange of risk-free germplasm;

iii) by developing sui generis methods of protecting the breeders' and farmers' rights by each country considering its own unique situation such that the benefits accruing thereof are also available to the other nations of the region; and

iv) by instituting funding mechanism and ensuring financial support for the regional PGR activities. It is pertinent to emphasise that the seed industry be facilitated and the security of certain vital components ensured.

Consequent upon the adoption of CBD and also implementation/enforcement of General Agreement on Tariffs and Trade (GATT)/ World Trade Organisation (WTO) provisions, new issues and challenges of global significance have opened up. Accordingly, the sovereign rights of the nations on their bio-resources have been specifically recognised. As a consequence, it has become responsibility of the nations to conserve these resources and ensure their availability for their systematic utilisation on sustainable basis.

Coupled with this, an equitable sharing of benefits accruing from the use of plant genetic resources (PGR) is an important issue that influences the other vital issue of access to PGR. The funding support from GEF has yet not set its pace which emphasises the need to set up a Global Fund which could inter alia help shape and refine the action points on seed security and the seed industry.

There is an urgency to address the identified areas in harmony with the other institutional mechanisms in the fray.

Recognising that the gene rich countries have a potential to release naturally evolved variability suited to several diverse situations, there is an adequate need to match with a generous global commitment towards the funding requirement to sustain the traditional/naturalised diversity on-farm at certain representative sites located within the diversity-rich areas of the world. Some activities might include, but not limited to:

i) Regional exploration teams comprising experts from countries involved could be assigned survey, collection and bio-prospecting of agro-biodiversity. Whereas the hitherto unexplored bio-resources should be collected, evaluated, documented and shared for immediate use besides long-term conservation in the gene bank, the well adapted and popular cultures should be identified along with their companion crops/varieties and the cropping systems, for their on-farm conservation;

ii) Seed multiplication, testing and distribution programmes should be launched for the traditional varieties/folk cultures/landraces which have edge over other co-cultivated variable forms and the farmers provided with critical inputs, either as subsidised or free-of-cost;

iii) Farmers should be encouraged to plant 'rest of the traditional cultures' in a small block of their cultivation field by ensuring seed of all the diverse component cultures from the agency/community seed bank. The produce should be re-purchased at the premium rate of seed rather than commodity;

iv) In order to develop regional capabilities/capacities, the human resource orientation and development needs to be given top priority, giving equal weight to educating both genders;

v) Existing gene bank facilities should be utilised for long term conservation of seed representative of diverse types for posterity. The satellite, medium term facilities, on the other side, should be used to maintain active forms with a view to replenish the lost or diminished types in a catastrophe/crisis; and

vi) Exchange of germplasm seed needs to be regulated by strict phytosanitary regulations and quarantine measures so as to minimise risk from new pathogens/pests prevailing outside the region. Joint ventures in bioprospecting, biosafety and biotechnology, including cryopreservation, in vitro conservation or handling of transgenics/genetically modified organisms (GMOs) need to dwell upon principles of sustainability and synergies/mutual co-operation. All such activities require adequate funding support which should be catalysed and facilitated as per need of the hour.

The damage being done to land, water, flora, fauna and the atmosphere altogether poses the challenge of future food security. It is the human interference that created the crop plants from the wild to their present day commercial forms. Their agro-biological features and dispersal mechanisms have changed drastically since then. The commercial forms are absolutely dependent on human beings for their perpetuation and survival. It is, therefore, appropriate to look into the interests of the seed producers of the required/desired types.

The management approach would require monitoring of the varietal component at frequent intervals so that the balance could be maintained between the interests of the seed producers and the phyto-ecological requirements of the varietal components. Involvement of both governmental and non-governmental bodies and the individuals, as also the genetic resource and patent literacy, constitute an essential elements of systematic sustainability and developmental plans.

Realisation in India and efforts on PGR Conservation

It is recognised that a balance between biodiversity conservation and economic growth continues to be an issue of concern. In this connection, the conservation and effective management of plant genetic resources is essential for achieving sustainable gains in productivity. Accordingly, there is a need to formulate and implement an Action Plan at the national and regional levels in conformity with the situations and provisions of the Global Plan of Action.

It is realised that the conflict of minds is universal. PGR were earlier held in trust as the 'common heritage of mankind' whereas recently the recognition of 'sovereign rights of nations over their biological resources' has come into effect. Accordingly, it is believed that the handling of issues related to seed security, distribution or replenishment in the areas of crisis needs more judicious treatment in conjunction with the agro-biodiversity related issues.

Agriculture at large, being the dominant sector contributing 31 per cent to gross domestic product (GDP) genuinely needs funding support for research, capacity building, infrastructure development, subsidy and grants, particularly in terms of attempting to maintain a guided composition of seed component over time and space. It has to simultaneously tackle the two contrasting dimensions of sustainability and development. Hence, a balanced, harmonised and optimised model of growth needs to be pursued.

The institutional mechanism provided globally by the CBD well recognised the need to harmonise with the International Undertaking on Plant Genetic Resources and its revision since 1993, in conjunction with the transformation in the purview of FAO Commission now re-designated as Commission on Genetic Resources for Food and Agriculture (CGRFA). The FAO's World Information on Early Warning System (WIEWS) clearly intends to safeguard the interests of agriculture through prospecting, country reports, replenishment of vital seed component in case of crisis, etc. in larger interest.

It is felt that agro-biodiversity be treated as a specialised discipline as a force to reckon with. The Conference of Parties of the Convention on Biological Diversity and the CG Centres/Global Forum have all shown concern that agro-biodiversity needs to be recognised specifically and should not be treated generally. Hence, the actions/issues, problems, potentials, opportunities, threats, etc., which needs attention could be listed as follows:

- The issue of growth is central to sustain the farm, family and livestock. This would essentially call for a balanced approach in conversion, use, cycle and recycle.
- As the conservation and effective management of plant genetic resources is essential for achieving sustainable gains in productivity, an action plan at the national level is being drawn to give due

emphasis on this component of biodiversity. The plan would be put to effective implementation during 1998.

- The tasks of PGR management would essentially have to be dealt with from two angles. The one in conjunction with other activities, and the other in collaboration with various stakeholders.
- Documentation of traditional wealth and knowledge of the country is well recognised. Documentation of traditional knowledge on all fields of plant wealth particularly on medicinal plants, skills, folk medicinal formulations, etc., is contemplated under a formulated programme to be implemented through a mission mode approach.
- The wild relatives of crop plants are the hope of future man-guided evolution of crop plants for posterity. The wild relatives of crop plants in India amount to over 325 species, which is a significant number and represents 1/5th of the total wild relatives' diversity available globally. There is a need to further study and exploit this rich treasure in the wilderness.
- A lot of national input has been made in raising a long-term collection of above 1.63 lakh seed samples of various crop plant species and their relatives in the National Gene Bank. This facility as a National Heritage is supported from the national resource for its sustainability and strengthening in order to ensure conservation for posterity.
- Other modes of conservation are also being supported on case-to-case basis. The tissue culture repository with a holding of nearly 1,000 cultures of about 30 different species together with the cryo-preservation facility holding over 2,500 accessions

of recalcitrant seed species on experimental basis, constitute part of the national gene bank. Field gene banks at different NBPGR stations have live trees of various horticultural plants, energy plantation and other economic tree species. Conservation of seed samples in the cold tracts at zero energy input is contemplated to supplement the collection at the National Gene Bank Repository. Research scale on-farm conservation activities are also envisaged and being pursued.

– Role of farmers in conservation of genetic diversity is well recognised. Farmers have to be convinced to practise on-farm conservation of their traditional cropping structures and system. This would give the world a hope to sustain and in getting naturally evolved forms in particular environments. For effective in situ on-farm conservation of genetic diversity, specific regions and species exhibiting concentration of diverse forms are being identified. It is believed that a system needs to be developed to find out the farmers' views and their expectations in sustaining in situ on-farm conservation of genetic diversity. Also suitable modes and mechanisms of incentives to farmers need to be developed so that the farmers resort to an on-farm conservation of the traditional types. Of the several options suggested, this may be done by expanding the Public Distribution System (PDS) to crops other than wheat and rice and also by chalking out market mechanisms for the traditional varieties of wheat and rice. Consideration should also be made to re-define productivity, so that not only yield but other components, such as, fodder (biomass) production, soil nutrient status, response to inputs, etc., are also accounted for and farmers get their due returns.

- To ensure sustained in situ conservation of agro-biodiversity, the areas/zones under traditional crops and continued traditional practices need to be earmarked and declared sensitive in situ on-farm conservation areas.
- There is need to develop instruments and provisions for Plant Variety Protection (Breeders' Rights), Farmers' Rights, Community Rights etc., in conjunction with effective linkages with the public sector, private sector, NGOs, Farmers' Associations and Communities for PGR conservation, in general, and collection and evaluation of germplasm, in particular.
- Emphasis has to be laid on human resource development to build up expertise in agro-biodiversity management, including legal implications and forms of access and benefit sharing. Germplasm collection is an expertise that would need stringent test over time when the depletion of resources further takes place globally. Germplasm management and conservation need equal priority. In this endeavour, the National Agricultural Technology Project is addressing the issue in right earnest.
- Recognising that information is a powerful tool and communication is the best means to generate awareness and to set the right track, India has launched a massive programme to address this issue in totality. Each district is provided on-line connectivity through electronic media.
- Preparation and dissemination of popular literature on various PGR related developments at the

national and the international levels, for creating awareness widely, including at the School, College and University levels is envisaged. Updating the information on intellectual property rights (IPR), CBD, PGR Acts and the development of software for literacy are considered necessary with adequate budget.

- Access to PGR among the contracting parties is a matter that is still receiving international debate. Benefit sharing from germplasm collection already existing in the CGIAR Institutes collected from developing nations is also being considered at length. The issue is under debate to reach to a logical conclusion.
- Legislative measures, such as follows, are being taken up at the national level to cover the interest of intellectual property protection:
  - i) Plant Variety Protection and Farmers' Act (almost at final stage);
  - ii) National Biodiversity Act (at drafting stage); and
  - iii) Patent Act (Revised) (being debated in the Rajya Sahba through a sub-committee of Parliament).
- A national level Advisory Committee on Genetic Resources is in the offing to give a sense of direction on all related issues.
- Memoranda of Understanding (MOU) and Material Transfer Agreements (MTA) need definite input to face all types of challenges. Conflicting issues related to PGR have to be taken up and harmonised in several brain storming sessions. A few such deliberations are on.

- A sound database on genetic wealth is a pre-requisite for an effective conservation strategy. A huge expenditure is estimated for the inventorisation and genetic resources conservation at the national level which has to be generated mainly from public sector. A National Fund is contemplated.
- Among the issues related to conservation of agro-biodiversity, we must give a national treatment and priority to the seed gene bank at NBPGR and also include, besides the identification of conservation sites, the conservation partners. We must decide upon as to which sites for in situ on-farm conservation to begin with. To identify genuine NGOs and other partners for their involvement in PGR Conservation is an important task. Feed back information from the NGOs and other stakeholders is essentially required as well as the dialogue with seed companies and other private sector people engaged in PGR activities. We should decide on the above issues genuinely and initiate action.
- Open attitude, dialogues and transparency are important for the perpetuity of the national efforts. A need for substantial financial support is felt.
- It has been planned to equip all the co-operating centres/data generating sites (NAGS) with microcomputers and necessary software. The National Information System has to cater the needs of all partners of Indian National Plant Genetic Resources System (IN-PGRS) including other countries. A thorough system analysis has been done in this respect in terms of input and output including their generation and standardisation. Due consideration is also given to information flow,

security/protection against misuse, hardware and software, manpower and communication facilities. Efforts for development of on-line system, availability of the system on the Internet are also under active consideration under the aegis of ICAR's Agricultural Research Information System (ARIS).

Efforts at the FAO Level

The salient points related to the progress of the global system on conservation and sustainable use of plant genetic resources for food and agriculture which emerged in the seventh regular session of the FAO Commission on Genetic Resources for Food and Agriculture (CGRFA) at Rome during 15-23 May, 1997 may be reviewed as follows:

- A multi-year work programme on agricultural biodiversity established as per decision III/11 of COP shall be pursued by the FAO Commission and CBD Secretariat though co-operation. Similarly, FAO's monitoring/reporting on agricultural biodiversity shall be consistent with CBD, other relevant inter-governmental bodies and world food summit' Plan of Action.
- CBD through its decision II/5 has requested GEF to give priority support to efforts for conservation and sustainable use of agro-biodiversity. Many countries stressed the need to have separate funding! FAO commission requested FAO to provide assistance and guidance to countries on request in implementation/ monitoring of the Global Plan of Action (GPA). Further, the monitoring of GPA and updating/reporting on the state of world's PGR were held complementary

activities which do not need separate reports (i.e., need only one report each) by countries.

- No changes in the CCGCT (Code of Conduct for Germplasm Collecting and Transfer) or any further work on Code of Conduct for Biotechnology were recommended, till the revision of IUPGR. However, a decision guide on regeneration of accessions in seed collections, prepared by IPGRI, has been made available to the FAO commission for comments by members.
- The existing arrangements between FAO and the 12 CGIAR centres shall be extended, pending the revision of IUPGR.
- Statutes of the inter-governmental technical working group on AGR (Animal Genetic Resources) and PGR were presented. Both working groups have 5 members each for Asia. The AGR were specifically addressed in view of the broadened mandate of the Commission and to facilitate integrated approach to agro-biodiversity.
- The summary statement presented by IPGRI informed the Commission that 0.5 million accession were now covered under the FAO-CGIAR agreements with the designation of 50,000 new accession under this arrangement.
- The summary statement presented by the CBD Secretariat showed that at least three decisions of the COP-III, Buenos Aires, Nov., 1996, were of direct relevance to the ongoing negotiation for revision of IUPGR, namely, Decision III/11—Conservation and Sustainable use of agro-biodiversity, Decision III/15—Access to Genetic Resources, and Decision III/17—Intellectual Property Rights. The COP-III noted the narrow

options for legal status of revised International Undertaking, as being,

i) a voluntary agreement,

ii) a binding instrument, and

iii) protocol to CBD.

It encouraged parties to implement GPA at national levels. Pollinators and soil-micrograms in agriculture needed special attention (as bio-indicators) and comprehensive reporting by countries to Subsidiary Body on Scientific, Technical and Technological Advice (SBSTTA)/ Conference of Parties (COP) of the CBD.

- An identified group of competent people needs to resolve the PGR issues and help prioritise various PGR activities. There is need to identify crops/ regions on priority.
- The dialogue between the public and private sector should continue and there is a need to create awareness about PGR.
- An information bank should be established to enhance literacy, management and utilisation of genetic wealth. Emphasis should also be given to regional languages in developing the information base.

World Information and Early Warning System (WIEWS)

The FAO's initiative to develop a systematic information base on PGR and related technologies under the World Information and Early Warning System on Plant Genetic Resources (WIEWS) is aimed to collect, collate, maintain and disseminate facts and figures on country profiles, national PGR programmes, ex situ collections, database on PGR, existing crop varieties, seed production and supplying agencies, in situ on farm conservation, bioprospecting and exchange of seed.

The nature of such an activity requires a comprehensive information input and a grid approach for analysis and interpretation of data. The system is open and responsive to queries from all stakeholders and interest groups. In practical terms, it would be relatively simpler to attend to details of the resource availability and the custodian(s) but, considering the importance of implementing CBD and WTO provisions and the imminent revision of the International Undertaking on Plant Genetic Resources, the access to resources for the purpose of replenishment could pose certain difficulties. The access would have to be required from any one of the following sources:

i) the ex situ seed collections;

ii) the seed multiplication and supplying agencies; and

iii) the local communities/community gene banks.

The implications, in each case, could be several and manifold. The access from ex situ CG system collections is being debated wherein two clear-cut options have been put forth. The pre-CBD collections are still available freely which means their replenishment would be only a matter of time actually required for the multiplication of their seed. However, in case of requirement of post CBD collections, several access-related issues shall have to be sorted prior to the multiplication of seed for replenishment.

The requirement of seed from commercial/public seed multiplication and supplying agencies could be easily met but such a replenishment would not serve the purpose effectively. This being so, the agencies could provide seed of few dominant varieties, which they are currently handling, in contrast with the requirement of a wider genetic base. The replenishment of seed from farmers' varieties/land races faces two different situations:

i) the issue of consent/access; and

ii) the issue of compensation or sharing of benefit accrued from the use of their heritage seed.

The setting up of community seed banks and the Global Biodiversity Fund have definitely been provided as suitable institutions which could help in build up at an appropriate system and come to the rescue in cases of crisis. The setting up of a separate Global Biodiversity Fund is still being debated as also the Farmers' Rights. The development of the sui generis systems to meet the countries' own (local) needs, has to be accelerated. Through this mechanism, the whole idea is to encourage local level replenishment through procurement of required seed from nearly/similar climate logical areas.

Policy on Seed Security for Effective Intervention

Primarily there are two broad issues:

i) Availability of seed for bringing much needed improvement to meet the existing and emerging needs of uninterrupted varietal improvement; and

ii) To have in hand seed for direct intervention and replenishment/utilisation for commercial cultivation to meet the normal and or contingent situations.

On the first issue, collection, conservation, maintenance, documentation, storage, mode, mechanism and system development for facilitated access with pros and cons on a balance sheet need to be put forth to assess various policy implications for exploitation of genetic resources on a long term sustainable basis. To deal with the second issue, concerted efforts on a long term perspective plan basis would also be urgently called for as follows:

1. As we go along to harness the productivity on a higher plane with the stipulation to attain the maxima; and the maxima being always dynamic in space and time; and with the localised increase in biotic and abiotic pressure in many of the location specific zones, tailor-made varieties to harness the potential of the eco-region and the system would essentially be called for. It would ultimately be the system's productivity on a balance-sheet which would provide enhanced productivity and profitability by balancing the input and output and the final turnover. In this endeavour, there would be a 'contradiction' as potentially the entire crop improvement programme rallies round conversion of solar energy by means of a biological system which is portable and marketable, be it food, feed, fodder, fuel or fossil.

   In order to attain efficiency of a high order of a plant system, the genetic tailoring at a narrow plane would obviously be the most crucial factor. This would hinder wide adaptability and thereby put a problem on stored seed utility to meet the contingent situation. If one needed to implement a seed security plan of a localised nature to meet the specific needs to harness higher potential, it would not be without a cost and obviously the cost would be of a high order as compared to the one in which a variety of a wider adaptability is considered. The cost would be lower in storing the seed of a widely adaptable variety as the manoeuvring and likely utility of the seed would be of a high order and, hence, the stock would be renewable and utilisable in a wider way.

2. To ensure that the tailor-made varieties are available, location specific breeding efforts would essentially be called for. At the same time, a finished product would need to be tested in different regions, situations and systems, be it of public or private origin. If of private origin, in particular, an instrument of plant variety protection would be imperative so that first rate technology at the global level could flow for the benefit of the farmers.

3. Provision of registration of varieties, system of varietal testing for their wider availability and suitability for cultivation cutting across countries and their documentation would be called for so that first hand knowledge is available as to what is suitable and for where?

4. A consensus and concerted efforts would be imperative for buffer stocking of seed. As a calculated risk and based on past experiences of the kind of calamities and eventualities, buffer stocking of seed to meet the national needs could be thought of. This would be an ex-ante decision as to what would be the kind of varieties and to what kind of situations so that only their seed is produced, stored and replenished. The private sector, keeping in view the market needs and likely commercial venture could also consider seed production and its storage and in turn need based marketing. It is quite likely that situation would demand for seed stocks to meet national calamities as has been witnessed in certain parts of the world, say in Ethiopia, where a good amount of seed was rushed

from neighbouring countries. At times, varieties were suitable and many times they were not at all congenial for commercial cultivation, but with little option.

5. It is a recognised fact that pest and pathogen move with the seed and in the liberalised world the problem of phytosanitary maintenance and post entry quarantine would be far more than ever before. As such, these regularity facilities would essentially be required to be strengthened particularly in case of seed borne diseases and viruses. It would be necessary that, in general, a common code of conduct is formulated and implemented after international consensus building. FAO can play a prominent role in this direction. Seed certification to ensure the quality of the seed would be of paramount significance. Efforts should be made so that strict provisions are adopted and implemented by various nations so that their certificate is globally recognised and supply of quality seed is ensured.

6. In the present world, the issue of free or open market has been discussed considerably. But one needs to precisely understand that it is not a free market or an open market or a liberalised or a globalised market. In fact, it is a competitive market in terms of quality and cost and it is the quality to which the global environment must suit so that the one which can bring added returns on investments made is allowed to flourish through seed culminating into food and feed security in the posterity.

7. Under the WTO, nations have agreed to enact a system of plant variety protection. The system must

be put in place so that a congenial environment is provided for greater investment in plant breeding efforts and consequent seed production and its dissemination. The protection by appreciating the intellectual property and ensuring returns on the investments made to some of the existing problems on quality seed production and its availability could be overcome. Hence, intellectual property protection as a means for ensuring added investment be appreciated and much needed incentive and reward system be created commensurate with the dimensions of the problems. Lest, seed security would be difficult to be ensured and hence, food and nutritional security would be jeopardised.

8. For ensuring much needed nutritional security at the household level through varietal constellation, it would be desirable that quality aspects of the produce are appreciated and the worth of the produce is provided in the market which could trigger quality related improvement programmes in the right direction.

9. The estimates at the global level has clearly brought out that to ensure food security, productivity of foodgrains will have to be doubled in the next 50-55 years. This would happen only if accelerated efforts on varietal development front is made and much needed insulation through resistance breeding programmes is put in place for various biotic and abiotic stresses at the earliest possible. It is recognised that apart from the varietal potential per se, the cropping intensity will have to be essentially enhanced where net monetary returns per unit area, input and time should be the

determining factor and system productivity in totality needs to be the determinant. This would call for a system wide system approach on varietal development, seed production, storage and its effective utilisation in a region, situation or system.

The primitive land racers or cultivars are virtually a blend of types, closely related. Many times, such mixtures of types provide greater insulation in production because of their inherent varietal blend situation. This kind of scenario is particularly effective in terms of moisture stress or salt stress or even in case of disease and pest problems where horizontal resistance is automatically brought in the fore front. There are very limited efforts on improving the potential of land races, although many of them could be of great consequence to meet effectively the local situations. In this endeavour, farmers' participatory approach to plant breeding is worth considering. But this approach could be adopted to regions, situations, crops and commodities where improved production technology has not been brought to bear on the productivity front due to improved varieties or monoculture practice of cultivation.

The production of identified promising types, their storage of seed and its use can go a long way in environmentally disadvantageous situations, be it drought prone or hilly regions or coastal situations. Hence, the issue needs concerted thinking, pragmatic approach and well-drawn policies and programmes to improve the existing land races by mere selection and following the pedigree method of breeding involving local people who would be the end-users.

6

Seed Sector and Food Security

Formal Seed Supply System

The formal seed supply system in the region originated from the common pattern that prevailed in centrally planned economies. The entire sector was directed and controlled by the state, from seed breeding to actual supply, and to the organisation and management of farm units. This explains the importance of that system and the fact that it is considered as the only one existing in the country. With transition, it became necessary to regulate the sector and the role of new parties through seed laws and operating regulations. Most countries in the region started to draft such legislation and to submit it to their respective parliaments for approval.

In most countries, the drafting and approval process was slow. Approved laws and regulations had to be modified at least once, in order to incorporate amendments that took into account the requirements of the market and of international schemes and rules (e.g. OECD, ISTA) to which the country had acceded meanwhile. The process did not always call for contributions or views from all parties in the seed sector; this generated criticism and may be a source of concern in the future. A number of countries received assistance from international organisations (e.g. FAO) for drafting or improving their seed law.

The legislative system in the Russian Federation with regard to agriculture in general, and to the seed sector in particular, is still not clear. Though, in theory, appropriate laws were promulgated and various decrees passed, seldom, if at all, have these legal instruments been enforced. The liberalisation of the internal seed market with the multiplication activity of imported seed required new legislation addressing the issue of Plant Breeder's Rights (PBR). Progress in this area has been very slow and this has delayed the full development of any large-scale private breeding activity in the region.

In all countries, the state has retained a regulatory role, assigning to government institutions the responsibility for testing and evaluating new varieties proposed by breeders and for registering them as improved cultivars in a national list, if they meet all prescribed requirements. In some countries, the institution is assisted by a consultative committee of independent experts. However, the final decision often remains with the Ministry of Agriculture or its equivalent, where appeals can be lodged in case of refusal.

The registration of an improved variety in countries like Poland also means that the seed is automatically released for multiplication. Where this is not the case, the date of the release and the quantity of seed to be multiplied have to be authorised separately. The inspection of seed producers' fields, control of the quality of seed marketed and inspection of processing plants remains the responsibility of the state, through an inspecting institution with some decentralised laboratories throughout the country (usually one per district or administrative region).

In some countries, in order to avoid conflicts of interest, this institution reports directly to the Ministry of Agriculture and not through the structure of the Ministry of Agriculture, in view of the Ministry's range of responsibilities in the formal seed sector. The supply of seed has generally ceased to be the responsibility of the state. Farmers have to use the market and privatised state enterprises where these have the monopoly of supply. However, in order to market and sell their seed, enterprises need an authorisation from the state and are subject to inspection by the designated institution.

In some countries, the purchase of improved seed has been subsidised by the state in order to ensure, through good yields, an adequate level of production. Those subsidies have sustained, indirectly, also the breeders and the producers of good seed. Countries that export seed must meet prescribed international standards, and so those countries have equipped their seed inspection and control institution with laboratories that can certify when those standards are met. Imported foreign seed must meet prescribed sanitary and quality standards; controls are carried out by customs authorities and by the relevant inspection institution when the product is marketed.

All governments in the region are aware of the importance of good quality seed in contributing to increased agricultural productivity and production. Therefore, the principal seed policy of governments of the region has been to ensure a continuous and wide use of high quality seeds. With the transition, the focus has moved to a less monopolistic role of the state and to establishing a basis for the private sector to develop. Nonetheless, factors internal to the seed sector, as well as external constraints, have required continuing public involvement.

Some countries have felt the need for, or have already established, advisory committees providing general policy guidance on issues concerning the formal seed sector. In countries like Poland, the seed trade association has taken the initiative to develop and propose a strategy for the entire sector. However, due to demands of structural adjustment programmes, most governments are at present cutting subsidies to the seed sector. As a result, government involvement in providing agricultural inputs, marketing, and research and extension services is being reduced. These changes have led to a greater emphasis on privatisation of the seed supply sector.

However, if not economically justified, private seed companies may not deliver seed to farmers living in marginal areas. National policies include full observance of the principles of the preservation of genetic resources, the maintenance of national germplasm collections and—to the extent possible—of breeding activity. This derives from the belief that agriculture was and remains a strategic sector. Also, the seed inspection and control function is firmly considered as a function to be assigned to the public sector. There is no indication that seed policies recognise or support the informal seed sector where it exists.

Breeding and Seed Research

Plant Breeding

In general, agricultural research activities have been centrally planned and entirely state financed, with research activities carried out by universities or similar institutions and by institutions that report directly to the Ministry of Agriculture. However, there are cases, such as in Hungary, where research institutions report to the Academy of Sciences. Universities focus more on the fundamental and theoretical issues, while other research institutions focus on applied research. With the transition and the serious lack of resources, the number of agricultural research institutions has declined in many countries, including the Czech Republic and Slovakia. Some of those that remain have become private (e.g. owned by the former staff), others survive with financial support from the Ministry of Agriculture and from income generated by their own activities.

In other countries, such as Kazakhstan, the agricultural institutes continue to belong to the state, but with minimal activities due to lack of funds. For the rest of the countries, the situation is intermediate, with no private institutes yet, but with a need to survive through the sale of products or services. In several countries, concern was expressed over the dwindling stock of seed laboratory materials and supplies, and the lack of resources to upgrade long-term cold storage chambers or the existing technology. So far, the public sector has supported the private sector mainly in seed production, marketing and trade activities, and, to a lesser extent, in the development of new varieties. The involvement of the private sector is essential for injecting fresh resources.

The primary agricultural research activities in most countries of the region involve variety development, testing throughout the country, release and registration, variety maintenance and breeder seed production. Plant breeding, the source of modern varieties, is the backbone of the formal seed sector. It used to be a government undertaking in all countries of the region, but now, with privatisation of the agricultural sector, first multiplication and then plant breeding increasingly have been undertaken by private or privatised enterprises.

Before the transition, the demand for seed of improved varieties was met by the state through its breeding institutions. As a result, there was no real need for PBR considerations. Now that the private seed sector has become involved in plant breeding, there has been a growing demand for the introduction of PBR. As a result, several countries have introduced appropriate legislation or are in the process of doing so. For all other countries, the matter is under active consideration.

The amount of hybrid seed and of vegetative propagating material imported by some countries (primarily for horticulture) suggests that their national research has not been as advanced and competitive as that of trading partners. The trend is already influencing the direction of future research programmes, focusing attention on the need to obtain practical and economically useful results.

Seed Improvement

Areas of seed improvement generally include: seed physiology, seed production, seed processing, seed quality control, seed health, seed storage and packing and some aspects of biotechnology. Following research, there is a need to circulate findings to the research sector and to communicate useful results to the education, agricultural extension and information sectors. Seed improvement programmes increasingly use biotechnology applications.

With a reasonable investment in laboratory facilities, biotechnology enables expansion of research and delivers faster results with a higher degree of precision. Biotechnology techniques are applied for virus and disease elimination, enhancing or modifying nutrient components, the conservation of germplasm, the development of biological pesticides, etc. There seems to be no substantial difficulties in terms of training technical and support staff, and educational and training programmes are being adapted to meet these needs in many countries.

Concern has been expressed that seed research financed by the state, including the use of biotechnology, is often remote from immediate application. As a result, findings from seed research are not bringing any income to the research sector. This is an important point to be considered in an overall strategy for the seed sector, as there must be programmes that reconcile the need for fundamental research (with long-term benefits) and applied research (with short- and medium-term benefits).

Certainly, research and training in business-like management of resources becomes indispensable if research institutions are to become partly or entirely self-financing. Research on genetically modified organism (GMO) applications exists in some countries, including Hungary, but only for studies related to food commodities. For the agricultural sector in general, and plant breeding in particular, the promulgation of legislation recognising intellectual property rights (e.g. PBR) is indispensable for further financing of research.

For the National Agricultural Research Systems (NARS) to be efficient would require funding for research, with full participation of all research institutes, agricultural universities, private sector firms, NGOs and farmers' associations of the country. It would also require the development of strategies and programmes, with the assignment of specific tasks and roles. International cooperation in the field of research already exists, but it is often restricted to a few countries (e.g. to a few universities in western Europe and North America, and to some traditional trading partners). This should be expanded to more countries of the region, particularly to those with common agroclimatic characteristics or constraints.

Most biotechnological research carried out at universities and through research centres are usually at a national scale. Interaction and regional cooperation between institutions and private industry dealing with biotechnologies could maximise resources and reduce costs. Furthermore, institutions already applying plant biotechnologies need to be involved in formulating national research policies and programmes.

Cooperation could also be further developed with some of the International Agricultural Research Centres (IARCs), such as the International Centre for Maize and Wheat Improvement (CIMMYT) for wheat and maize; the International Potato Centre (CIP) for potato; and the International Centre for Agricultural Research in the Dry Areas (ICARDA) for wheat, chickpea and lentil. However, the major problem confronting most IARCs has been declining and unstable funding levels.

Growing uncertainty regarding the stability of donor funding was highlighted at the World Food Summit, and is recognised as one of the major constraints to future food security throughout the world. In this regard, national and regional political and financial commitments to investment in agricultural research have been recommended.

Improved Seeds

Reports from most countries mention a large use of improved seed for a long list of crops and cultivars. In many cases, the percentage reaches 100%. Even though this percentage appears doubtful, it confirms that farmers prefer to use (and do use) certified seed when they can. It is part of their normal custom. To promote the use of improved seed, some countries of the region, including the Czech Republic, Slovakia and Hungary, provide subsidies to farmers to buy certified seed.

Due to some cases of trade malpractice, some farmers have been able to buy breeder seed or certified seed labelled as grain. It has often been said that due to the economic status of resource-poor farmers in the region, the cost of seed of improved varieties discourages their use. The statement is not entirely correct because the price differential between grain and improved seed is usually low in this region. However, the main reason for low use of improved varieties in this region is because improved varieties often require more agricultural inputs (fertilisers, herbicides, pesticides), irrigation water and handling than farmers can afford.

Research for developing improved varieties requiring less inputs would certainly be one means to improve food production and food security. A needs assessment of limited-resource farmers is necessary before new varieties are developed and introduced. Such a study could provide information on the needs and resources of the farmer and the climatic and edaphic conditions of a particular target area. The survey could also guide policy-makers in their decisions, so as to favour the on-farm production and utilisation of improved varieties by farmers.

Seed Production and Supply Systems

Seed production in the formal seed supply systems of the region does not differ significantly from country to country. In general, seed production requires a sequential process, which is vertically organised. It starts with Pre-basic seed, issued from Breeder seed supplied by a national or international research system. The Pre-basic seed is increased to Basic seed, from which several generations of Certified seed is produced. In some cases, uncertified, so-called "commercial" or "improved" seed, is produced and sold to farmers.

Throughout the region, the nomenclature of seed generations is fairly similar. The example below from Poland is typical of the general system in use. In Poland, for cereals and fibre plants, Pre-basic seed comprises Super elite and Elite; for fodder plants and small-seed legumes, it comprises Super elite and Breeding elite; for fodder grasses and oil plants, it is just Breeding elite. Basic seed is generally named "Original" throughout the region. Certified seed has two levels, called Certificate I and II. However, oil plants have only Certificate I, and beet hybrids have just the hybrid denomination.

There are regulations and procedures to be followed from one multiplication cycle to the next, and technical competence of the seed industry staff is fundamental throughout the production cycle. Seed producers are required to have a good knowledge of the crop and its phenotypic and genotypic traits. Seed produced by the formal seed sector is required to go through several testing procedures in order to ensure that quality is maintained over time. Seed legislation stipulates quality standards for each crop.

The implementation of testing, field inspection and quality control procedures gives credibility to the product marketed. Control is extended to processing plants and storage structures. Improved seed is required to go through a series of seed handling processes before it reaches the growers. These processes, which include seed processing, distribution, and marketing, require a certain level of seed standards. Throughout the region, seed processing facilities are usually adequate and decentralised.

There have been cases of malfunctioning of processing plants, of contamination from different-source seeds, and a breakdown in the seed supply system due to transport problems. The last-named becomes serious in case of shortages or prohibitive prices of fuel. Adequate management is needed to successfully operate processing plants, and training of technicians is crucial across the region if the seed produced by the formal sector is to reach the expected standards.

Another vital aspect of the seed industry is the function of conditioning and storage, which must be adequate to preserve the physical characteristics of the seed specified by regulatory standards. For storing certified and lower grade seed, less rigid principles need to be followed. Seed should be naturally or artificially dried to the moisture content recommended for each variety, and stored in ventilated, low-humidity facilities. These parameters, if met properly, are adequate for short-term storage. There are also locations that are perfect for storage due to optimal climatic conditions; however, if the distance between the storage area and the fields is considerable, the high transportation costs may suggest storage in less perfect areas or facilities.

Due to the general reduction in agricultural production and the import of seed and vegetative propagated material since the early 1990s, there has been overcapacity in the seed processing and storage plants. The main problems confronted by seed distribution systems are related to storage, transport and handling of the seed. In many countries of the region, seed marketing is also a weak link in the seed-production chain, limiting farmers' access to seed. At the same time, private seed industries, which have a business approach to seed production and distribution, have improved the seed marketing network in CEEC.

Marketing of imported vegetable seed is a well organised operation of the private sector, for example, in the Czech Republic and Slovakia. It is beneficial for countries that are striving to increase their activities in international seed trade and have not yet done so, to become effectively involved in international seed organisations, including the International Seed Testing Association (ISTA), the Organisation for Economic Co-operation and Development (OECD), the International Seed Trade Federation (FIS), and the Union for the Protection of New Varieties of Plants (UPOV). Lack of cooperation with these organisations decreases opportunities for export of seeds, thereby limiting the needed private sector involvement in the seed supply sector.

Seed and Emergency Situations

The large number of IDPs and refugees, the estimated size of the undernourished population in some 15 countries, and the projected increase in the population in seven countries (mostly from the same group) are exerting some pressure on the agriculture sector to improve and provide additional food. One of the answers is through the increased use of improved seed varieties. There have been situations where the supply of seed was interrupted, such as during large-scale civil disturbance and war.

In recent years, some countries in the region have undergone food shortages brought about by weather conditions, as well as complex internal strife and war, forcing the governments to use their foreign exchange resources to purchase food. Such situations place additional strains on economies and can cause tremendous economic setbacks. This situation has received the attention of several governments, the donor community, and international agencies (e.g. WFP). Seed is one of the primary needs of farmers displaced because of war and natural calamities.

Formerly, most affected countries received seed aid based mainly on humanitarian grounds, often with little consideration of the seed variety and its quality aspects. While such assistance has benefited some countries in the short term, especially in easing food shortages, the poor performance of untested seed material, and severe genetic contamination or complete displacement of local landraces and farmers' varieties, can cause disruption of the informal seed sector and damage the traditional production systems of a country.

Now, NGOs and other aid agencies pay close attention when purchasing food grain from neighbouring countries and distributing the grain to farmers. Failure to do so may introduce new problems of pests, diseases and weeds and cause genetic contamination and displacement of traditional varieties. Since some farmers in the region save a portion of their harvest as seed for the following year, this problem may continue long after the emergency and jeopardise future national agricultural development and food security. Clearly, coordination is needed between emergency relief and long-term development assistance programmes to avoid problems such as introduction of inappropriate crop varieties and pests.

National, regional and global strategies are essential to cope with emergency situations. If an area in a country is affected, the rest of the country can supply the required seed, provided that national seed programmes are well established. However, if the whole country is affected, the crop varieties can be supplied through other countries in the region or through regional IARCs and international development agencies such as FAO, WFP and UNDP, and NGOs. Such assistance can help the country resume its seed production without serious threat to the production programme.

When seed programme initiatives are implemented in harsh environments, it may be difficult to reach target population if unexpected calamities occur. Adequate safeguard measures often cannot be made in time to protect the seed and planting material. In order to avoid these risks, some remedial measures should be taken:

i) areas with some assurance of irrigation or adequate rainfall should be selected and farmers from such areas trained to produce seed;

ii) the quest for short-season and drought-resistant varieties should be the focus of plant breeders, using the genetic material from already existing landraces and other appropriate sources of (disease and pest resistant) germplasm;

iii) as has been done in some Asian countries, farmers should be kept informed through radio and other mass communication systems giving early warning of sudden weather fluctuations. Mass media can also promulgate advice on crop management;

iv) with investments and support from rural development organisations, every possible water source should be developed appropriately in a sustainable manner;

v) farmer groups should receive training on a regular basis, to assure the sustainable on-farm production of high quality seeds; and

vi) local banks and other lending organisations should be approached to provide credit as an incentive to farmer groups interested in on-farm seed production.

In most emergency interventions of the past, responses were ad hoc and voluntary. There was no standing capacity to respond appropriately to such disasters; no clear description of responsibilities, and no coordinating mechanism to bring the various agencies and organisation together for planning and implementation of emergency seed supply during and following disasters. This increased the risk of supplying inappropriate seed and planting materials and jeopardised the sustainability of future agricultural systems.

In response to these difficulties, the Fourth International Conference on Plant genetic Resources, held in Leipzig, gave FAO the mandate to coordinate programmes for rapid acquisition and multiplication, restoration and provision of seed to countries in need. This work is to be done in collaboration with WFP, the United Nations High Commission for Refugees (UNHCR), the Office of the United Nations Disaster Relief Co-ordinator (UNDRO), the International Plant Genetic Resources Institute (IPGRI), NARS, IARCs, regional PGR networks, governments of the countries affected, donor countries and NGOs.

To become food self-sufficient, farmers should have on-going access to quality seed in normal and crisis situations. Viable seed supply systems to multiply and distribute the seeds or planting materials that have been developed or preserved are critical for the success of food security programmes in the region. Therefore, seed security policies and programmes are needed in the region in order to assure a ready seed supply in the quantity and quality needed of acceptable and suitable varieties, and that are affordable to farmers. In one country of the region, news circulated that hoarding of grain by farmers in 1999 amounted to 5 million tonnes.

Informal Seed Supply Systems

Informal seed supply systems should not be confused with "black market" situations that exist in several countries of the region and affect both the formal and the informal systems. The informal seed supply system consists of farmer-managed seed production activities and is based on indigenous knowledge and local diffusion mechanisms. It includes methods such as retaining seed on-farm from previous harvests to plant the following season, and farmer-to-farmer seed exchange networks. There has been little or no emphasis on the informal seed supply sector and little is known about its operation in the region. As a result, documentation with regard to the informal seed sector is scarce.

At the time of centrally planned economies, the only seed supplier was the state, with its specialised institutions, and there was hardly any room for an informal seed sector to develop. However, it is conceivable that in countries like Poland, that always had a large number of small private farms, there was an informal seed system limited to the retention of good seed, processing and exchange within the same farmer community. A true breeding activity aiming at the development of improved seed varieties, and further activities like large-scale marketing, did not take place.

With the transition, the state ceased to be a supplier of low-price certified seed and of other agricultural inputs. With land privatisation, the number of small farms increased considerably, so small-scale farmers found themselves with limited financial means and with the need to economise on the purchase of seed and other inputs. Seed started to be replaced after a longer period, there were fewer purchases of certified or commercial seed, and, in extreme situations, farmers abandoned the formal seed market.

In some countries, the quantity of certified seed sold dropped by almost half. It would be useful to analyse the situation in more detail in order to reflect these developments in the comprehensive assessment of the seed sector of the region. The experience from other regions suggests that where an informal seed supply sector is not negligible, it should be included in the policy and programmes for increasing the use of good quality seed. An effective on-farm seed production system would require that:

i) surveys be conducted to identify the biological, social, and economic factors of the varieties important to farmers;

ii) breeders collaborate closely with farmers on such seed production programmes;

iii) local germplasm be collected for long-term storage in genebanks for future multiplication and use in production; and

iv) seed legislation take into account the low level of technology and minimum quality control procedures that are practised at the farmer-level. Assistance by FAO to governments in the region may be needed for the development of legislation.

In an informal seed supply system, seed distribution operates mainly, but not exclusively, between households within communities. However, lines of supply can be extended over a relatively wide geographical area. Under this type of seed distribution scheme, a wide variety of exchange mechanisms are commonly used to transfer seed between individuals and households, including cash sales, barter and transfers based on social obligations. The quantities of seed exchanged in the informal seed supply systems are often very small compared to amounts traded in the formal sector.

Regional Seed Supply Systems

Research, extension, input supply services, distribution and marketing are some of the major component of an effective seed supply system. The neglect of any one component in the seed development chain affects the entire seed supply system. Furthermore, changes in policies affecting one component may have adverse effects on the performance of others and jeopardise the development of an emerging seed supply system. Some examples of what may result when linkages are not adequate are:

i) new varieties may be developed by agricultural research institutions, but do not have the attributes required by farmers;

ii) distribution channels for improved seed may have suitable varieties available but fail to reach small-scale farmers in a timely manner, at an affordable price, or at accessible locations;

iii) distribution channels for complementary inputs may be ineffective; and

iv) extension services or agricultural credit institutions may fail to provide the necessary support to farmers.

Any of the above factors could limit adoption of improved cultivars, thus affecting the development of a seed supply sector. Linkages in seed supply systems must be considered when defining appropriate seed strategies a country and the socio-economic circumstances of its farmers. In addition, it is important to be aware of potential negative repercussions to the seed supply sector if changes in other policies affect services offered to farmers.

7

Seed Storage

When storing seeds for the future, be aware that different types of seeds should be stored under different conditions. Their ability to produce a plant depends on the time of seed harvest and the conditions of storage. Most seeds should be stored in a sealed, airtight container and kept in a cool, dark area. Mark the container with indelible ink including the plant and variety name, the date stored and other identification information. There are three categories of seed storage based on the life of the seed.

A viable seed is one that remains alive and is able to produce a live plant. Short-lived seeds are those that are viable for short periods, such as a few days or months under the right conditions. Medium-lived seeds are viable for two or three years if stored at low humidity and low temperatures. Long-lived seeds have hard seed coats. They are impermeable to water and can remain viable for up to 15 or 20 years.

To determine if stored seeds are still viable, take few of them, count them and place them in a pot or flat tray filled with potting mix. Water the seeds well and give them plenty of light. Keep the potting mix moist, and see how many seeds sprout. If more than half to three quarters of the seeds sprout, then the seeds have a good chance of germinating in the Garden. Hybridized seeds, which are developed from a complex genetic process, are grown and sold for superior characteristics like large flowers and very tasty fruit. Seeds collected from these plants will seldom produce similar desirable plants from their seeds. Do not save or store these seeds; in most cases, the resulting plants will be a disappointment.

The storage facilities to be provided must be related to the amount of seeds and the period over which they are to be stored. It is a waste of money to create expensive facilities capable of maintaining viability for 10 years if the seeds will never stay longer than nine months between collection and sowing. It is equally wasteful to spend money on seed collection, extraction and cleaning if storage conditions are so inadequate that the seeds are 90% dead before they reach the nursery.

For general accounts of seed storage of forest tree seeds, there are a number of useful references. They deal mainly or exclusively with temperate species. More intensive studies have been made of the storage of agricultural seeds and there is good reason to accept that the general principles established for agricultural crops apply also to forest trees. An excellent recent account of the subject as applied to agricultural seeds is contained in "Principles and practices of seed storage" and there is also much useful information in the slightly older publications of Roberts and Harrington. Long-term storage for gene resource conservation is well covered by Cromarty et al.

The period for which seed can remain viable without germinating is greatly affected by its quality at the time of collection, its treatment between collection and storage and the conditions in which it is stored. Nevertheless, seed longevity varies enormously from species to species even if they are given identical treatment and storage conditions. Ewart divided seeds into three biological classes according to the time for which they are capable of retaining viability under "good" storage conditions:

– *Microbiotic*: seed life span not exceeding 3 years
– *Mesobiotic*: seed life span from 3 to 15 years
– *Macrobiotic*: seed life span from 15 to over 100 years.

Although Ewart's classes were useful in drawing attention to the differences in natural longevity of seeds of different species, his classification is too rigid to fit the variations between individuals, provenances and seed years in a single species, or the possible variations in storage conditions. It is not possible to define a standard set of "good" storage conditions equally suitable to all species, because species vary in their requirements for optimum conditions. Yet storage life of a given species will vary greatly according to the conditions in which it is stored. Today two major classes of seed are recognised:

1. *Orthodox*. Seeds which can be dried down to a low MC of around 5% (wet basis) and successfully stored at low or sub-freezing temperatures for long periods.
2. *Recalcitrant*. Seeds which cannot survive drying below a relatively high moisture content (often in the range 20–50% wet basis) and which cannot be successfully stored for long periods.

Within these two classes some further subdivision may be made, for example between orthodox seeds with or without hard coats and between recalcitrant seeds which can or cannot withstand low temperatures of below around 10°C. Within each of the main classes there are still considerable differences between species in the period for which viability is maintained under a given set of conditions. There may also be a distinction between truly recalcitrant species and species that are just difficult; the latter may turn out to behave in an orthodox manner if, for example, particular attention is paid to the methods of drying them.

Orthodox Seeds with Hard-coats

Most, if not all, of the species which have been recorded as maintaining seed viability over a period of decades are hard-seeded. They include a number of tropical leguminous species. Examples of species of which at least some seeds maintained viability after lengthy periods of storage in herbaria, cited by Harrington from the work of Ewart and Becquerel, are:

158 years	*Cassia multijuga*
149 years	*Albizzia julibrissin*
115 years	*Cassia bicapsularis*
99 years	*Leucaena leucocephala*

Ambient conditions in herbaria storage can be considered good (fairly low relative humidity and temperature) but well short of the combination of low initial MC, sealed storage and sub-freezing temperature now considered ideal for longterm storage of orthodox species. Recent research has provided more precise information on conditions of storage, initial germination and final germination of some species, but over shorter periods.

Modern thinking has defined low MC, low temperature and low oxygen pressure as the three most important constituents of the storage conditions which man should provide to maximise seed longevity in orthodox species. In the impermeable seedcoat nature has provided two of these constituents, a low MC and exclusion of oxygen. Full-sized but green leguminous seeds, sown immediately without drying, may germinate at once, indicating that the seedcoat has not yet developed an impermeable layer; no doubt the development of impermeability is synchronised in nature with the reduction of seed moisture by natural drying to the optimum content for longevity.

Hard seeds are thus a potent factor in extending seed life in all conditions of storage but confer their most important benefits when storage facilities are limited and during the potentially dangerous period between collection and entry into long-term storage. Not all leguminous seeds are equally long-lived, for example *Koompassia malaccensis* seeds have thinner seedcoats and deteriorate more rapidly in storage than species such as *Parkia javanica,* and they need no pretreatment to overcome seedcoat dormancy.

In Sudan seeds of *Dalbergia sissoo* stored less well at room temperature than those of local *Acacia, Albizzia* and *Tamarindus* species, while in Australia seeds of *Acacia harpophylla* deteriorate rapidly unless stored in sealed containers at 2–4°C.

Orthodox Seeds without Hard Seedcoats

Many species in important genera of forest trees fall into this group, e.g. *Pinus, Picea, Eucalyptus*. Experience in Australia is that mature seeds of all eucalypts can be kept viable for some years if stored with a low moisture content in sealed containers at 3°- 5°C. The majority of species can be stored for 10 years at room temperature with relatively little loss of viability. Both *E. deglupta* and *E. microtheca* seeds deteriorate more rapidly if stored at room temperature, but storage life is improved if they are kept in air-tight containers at 3°- 5°C and recent evidence suggests that storage at -18°C is even better.

In Thailand seeds of *P. kesiya* and *P. merkusii* retained good viability for four years if stored at below 8% moisture content in sealed containers at 0°- 5°C, while at least five years' good viability is possible with *P. caribaea* and *P. oocarpa* under similar conditions. Considerably longer periods have been recorded for some species of pine e.g. 30 years for *Pinus resinosa* in USA when stored in sealed containers at 1.1°- 2.2°C. *Tectona grandis* is an orthodox tropical broadleaved species but, since it produces good seed crops in most years, there has been little stimulus to investigate optimum conditions for long-term storage.

According to evidence summarised by Bowen and Whitmore, most *Agathis* spp. are orthodox. For example, an appropriate treatment of *A. australis* in one study (dried to 6% MC, then stored in sealed containers at 5°C) preserved viability for 6 years (79% germination compared with the initial 88%), while storage at below freezing temperature maintained a germination of about 60% for up to 12 years. The same seed stored at higher MC or temperatures (15 - 20% MC. or 15°- 20°C) had lost all germination power within 14 months.

A australis seed is inherently longer lived than *A. robusta* which in turn is longer lived than *A. macrophylla.* Initial trials with the tropical *A. macrophylla* indicated that good results could be obtained by drying fresh seeds from about 65% to 20% before despatch by air (period in transit 14 days) and by further drying in the recipient country for 5 days at 16°C and 14% RH. final MC was 6% and germination 75%. However, later trials were inconsistent and less successful. With tropical species, it is likely that handling between collection and despatch and the largely uncontrollable conditions of air transit are more critical than in the easier temperate or subtropical species.

Orthodox species which rapidly lose viability unless they are given the optimum treatment include species in the mainly temperate genera *Populus, Salix* and *Ulmus.* Many of these lose viability within a few weeks under natural conditions or if stored in ambient conditions of temperature and humidity, but can be stored for months or years if maintained at low temperature and low moisture content. Examples are of *Ulmus americana* stored successfully for 15 years at 3% moisture content and -4°C, and *Populus sieboldii* stored for 6 years at -15°C over a desiccating agent in a sealed container.

In *Populus balsamifera* and *Salix glauca,* reduction in germination after two years of sealed storage at -10°C was less than 6.5% of initial germination; after three years there was very little change in *Populus* but up to 40% reduction in *Salix.* In the tropics *Aucoumea klaineana* is a good example of an orthodox species which is short-lived under ambient conditions. Germination of fresh seed is often over 90%, but after 30 days' storage in room conditions there is a significant drop in germination and this falls to zero after 100 days.

Storage at 0 - 5°C and 7 - 8% MC in sealed containers with a chemical desiccator, Actigel, maintains over 50% germination for at least 30 months. There is some indication that a further reduction of MC will conserve viability even better. Thus one seed lot of initial germination 76% in the laboratory and 79% in sand, when stored in sealed containers with Actigel, had an MC of 4.6% and a germination of 70% in the laboratory and 79% in sand after 30 months; the same seed lot stored in sealed containers without Actigel had an MC of 9.9 -

10.4% and a germination of 54 - 63% in the laboratory and 62 - 67% in sand.

Other species of this type include *Entandrophragma angolense* which has a seed life of 6 weeks in room conditions but up to 6 years in cold storage and *Cedrela odorata* which loses all germination capacity in 10 months at room temperature but suffers no loss in 14 months if stored at 5°C in a sealed jar. Some species may need special treatment to prolong viability for more than a few months. *Fagus sylvatica* can be conserved overwinter by maintaining MC at 20 - 30% and storing part-filled in sealed polythene bags at 0 - 5°C for 100 days. It is then suitable for sowing because such storage conditions constitute a suitable pretreatment to break dormancy.

If longer storage is intended, MC should be reduced to 8–10% by drying in a current of air at room temperature (15–20°C). The nuts are then placed in sealed containers and stored at -5°to -10°C and will keep for several years. Later research in France and Poland confirmed the above MC of 8–10% and the advantages of sealed containers for long term storage. This technique has been successfully applied on a large scale (17 tons of beechnuts from 51 different sources) in France. Germination has been maintained over periods of 4 to 6 years.

Where storage conditions leave much to be desired, the longevity of orthodox seeds without hard coats can be expected to be much inferior to the hard-coated species. The nearer that conditions of storage approach the ideal for a given non-hardseeded species, the less the difference between its longevity and that of a hardseeded species. The best combination of MC and temperature will vary to some extent between species, for example the above-quoted 8–10% MC for *Fagus sylvatica* is considerably higher than the 5-6% considered ideal for many forest and agricultural seeds.

Recalcitrant Seeds

Recalcitrant seeds include a number of large seeds that cannot withstand appreciable drying without injury; it is of interest that the overwhelming majority of recalcitrant species listed by King and Roberts are woody. Temperate species such as *Quercus* and *Castanea* are commonly stored moist only for short periods over winter. Reduction of storage temperature to near freezing will prolong longevity. Bonner a) found that it was possible to store acorns of *Quercus falcata* for 30 months and still obtain over 90% germination at the end of the period, provided that temperature was maintained at 3°C and MC between 33% (initial) and 37% (final). A lower MC or a higher temperature (8°C) both reduced germination.

For *Quercus robur* MC should be maintained above 40%. Recent research in Poland has demonstrated good results from storing this species at >40% MC in air-dry peat or air-dry sawdust in milk cans at -1°C. It is important to allow free entry of oxygen and this is ensured by inserting several strips of cardboard at intervals between the lid and the edge of the can. In these conditions germination after 3 winters was in the range of 38 - 75% and after 5 winters was still about 12%.

Temperatures below -5°C killed all the acorns, while a temperature of +1°C encouraged excessive pregermination (60 - 75% after 3 winters, with radicles up to 25 cm long, compared with 12% and radicles <0.5 cm long at -1°C). There may be possibilities of storing seeds after emergence of radicles. Recent research in Poland has indicated that best results are obtained with the recalcitrant *Acer saccharinum* by maintaining MC at the same percentage (50–52%) as when the seeds were freshly collected. For *A. pseudoplatanus* in the UK a minimum MC of 35% is recommended, while in Poland an MC of 24–32% and a temperature of -3°C have proved suitable to store samaras over three winters.

Most short-lived recalcitrant tropical species are constituents of the moist tropical forests, where conditions conducive to immediate germination (high humidity and high temperature) are prevalent throughout the year. Typical genera are *Hevea, Swietenia, Terminalia* and *Triplochiton,* as well as a number of Dipterocarp genera such as *Dryabalanops, Dipterocarpus* and *Shorea* and some species of *Araucaria. Dryabalanops* is injured if dried below 35% moisture content (MC) but still survives only about three weeks at over 35% MC. *Triplochiton* seed is naturally short-lived but can be stored for up to 22 months at a temperature of around 6°C and a moisture content of between 12 and 25%.

Azadirachta indica seeds also have a short period of viability, although the species occurs in dry, not moist, tropical forests and it is not clear whether it is a genuine recalcitrant or simply a short-lived orthodox species. Orthodox and recalcitrant species sometimes occur within the same genus. In *Acer* and *Ulmus,* genera in which both orthodox and recalcitrant seed behaviour occur, the distinction in North American species is clearly between spring- and fallseeders. *A. rubrum* and *A. saccharinum* flower and seed in the spring. Their seeds are not dormant, and their storage behaviour is clearly recalcitrant. Other *Acer* species have fall-maturing seeds, which are dormant and orthodox in nature at maturity.

The same occurs in *Ulmus.* Seeds of *U. crassifolia* and *U. serotina* mature in the fall, and are orthodox in storage behaviour. Spring-seeding species of *Ulmus* are "weakly" recalcitrant. In *Araucaria, A. cunninghamii* and other spp. in the Eutacta taxonomic group behave as orthodox. In Queensland seeds of *A. cunninghamii* of 5 provenances were air dried and stored at varying temperatures in sealed and unsealed containers.

At the higher temperatures, +1.7°C and -3.9°C, germination started to drop after 17 months' storage and after 8 years was down to about half the initial germination rate in sealed containers and about one third in unsealed containers. At the lower temperatures of -9.4°C and -15°C, germination after 8 years' storage was little changed from the initial (41–44% compared with initial 49%), and there was virtually no difference between sealed and unsealed containers. The rate of viability loss at the higher storage temperatures varied from provenance to provenance, but all stored better at the lower temperatures.

Moisture content was not recorded but under local conditions air-dry seed is normally in the range of 16–23%. Later trials with Papua New Guinea *A. cunninghamii* have shown that seeds can be dried from 21% to 7% MC without any effect on initial germination rate; effects on storage life are still under investigation. *A. hunsteinii* in the Intermedia group and *A. angustifolia, A. araucana* and *A. bidwillii* in the Colymbea group are apparently recalcitrant. Arentz found that high viability of *A. hunsteinii* could be maintained for at least 6 months by storage at 3.5°C and high MC; 37% was significantly better than 32%.

For some temperate recalcitrant species a relatively low temperature (just above or just below 0°C) has been found beneficial in extending the life of the seeds; low temperature to some extent compensates for the high MC which must be maintained to prevent the early loss of viability. In some tropical species, seeds are quickly killed if temperature is reduced too low, just as they are quickly killed if MC is reduced too low. Among woody species cited in King and Roberts are *Theobroma cacao* (killed below +10°C), *Mangifera indica* (damaged below +3°to +6°C) and, among the dipterocarps, *Hopea helferi, Hopea odorata* and *Shorea ovalis* (damaged below, respectively, +5°C, +10°C and +15°C). This susceptibility to chilling damage at temperatures above 0°C compounds the difficulty of storing these recalcitrant species, which seldom maintain viability for more than a few weeks or at most months. This compares with a normal seeding periodicity of several years in most dipterocarps, so there is no possibility as yet of conserving seeds in a viable condition from one good seed year to the next.

Unlike orthodox species, in which viability is best preserved by maintaining a minimal respiration rate, it appears that active respiration is necessary to survival of seeds of most recalcitrants. Thus damage to recalcitrant seeds has been reported not only from inadequate MC and too low a temperature but also from lack of oxygen e.g. in *Araucaria hunsteinii, Hevea brasiliensis* and *Quercus* spp. Whereas some temperate recalcitrant species have been stored successfully for several years, seed longevity in tropical recalcitrants can be measured in days or weeks.

The amount of research on tropical species is still small, especially on forest species, and it is possible that seed longevity could be prolonged beyond a few weeks if the best combination of seed maturity, speed, conditions and degree of drying, and most suitable storage temperature could be determined for each species. King and Roberts suggest a research strategy.

Effects of Factors on Longevity

Seed Condition

Even in ideal storage conditions seed will soon lose viability if it is defective from the start. Factors to be considered are:

Seed maturity: Fully ripened seeds retain viability longer than seeds collected when immature. Certain biochemical compounds, essential for preserving viability, may not be formed until the final stages of seed ripening. These include dormancy-inducing compounds in certain species, and dormancy is sometimes associated with seed longevity. In a few species e.g. *Gingko biloba, Fraxinus excelsior,* seed embryos are underdeveloped when the seed is shed. Maturation of these embryos is necessary before sowing but need not be done before storage. In *Fraxinus excelsior* drying of freshly collected samaras to 9–10% MC, followed by storage in sealed containers at -3°C, gives satisfactory results provided that successive moist warm and moist cold treatments are applied after storage.

Parental and annual effects: In seed harvest, quantity and quality often go together. The percentage of sound seeds in a high-yielding mother tree is usually higher than in one with a poor crop. Similarly, a given mother tree will have a higher percentage of sound seeds in a good crop year than in a poor one. Collection from high-cropping mother trees in a seed year is likely to yield seeds with the best longevity in storage. On the other hand, high-cropping "wolf" trees should be avoided because of their potentially undesirable wood properties, even though they may produce long-storing seeds.

Freedom from mechanical damage: Seeds damaged mechanically in extraction, cleaning, dewinging etc. rapidly lose viability. The danger is greatest for species which have thin or soft seedcoats. Excessive heat during extraction or drying also damages seed. Care should be taken to use the minimum times, lowest temperatures and minimum machine speeds necessary during the preparation of seed for storage. In some species, damage during dewinging may be reduced by partly restoring the moisture content between extraction from cones and dewinging, since moist seeds suffer less mechanical damage than dry ones.

Freedom from physiological deterioration: Poor handling in the forest, during transit or during processing causes physiological deterioration of seeds even if mechanical and fungal damage are absent. Adequate ventilation of orthodox seeds is necessary to avoid rapid respiration and overheating, while recalcitrant seeds must be protected against excessive drying.

Freedom from fungi and insects: For species stored at low temperature and low moisture content, the storage conditions themselves should prevent the development of fungi and insects. It is necessary, however, to avoid collection of crops showing a high incidence of fungal or insect attack and to carry out all operations of collection, transport, processing etc. as quickly as possible to ensure seed is not already damaged before it goes into storage. Attack by fungi and insects is most rapid on the forest floor, so collection from the ground should be carried out as soon after fruit fall as possible.

Fungicidal treatment cannot be generally recommended since it can be harmful to seeds; many fungicides are only effective when dissolved in water and are inappropriate for dry storage. Insects are usually killed if seeds are dried at temperatures above 40°– 42°C. For seeds which cannot be dried, other measures may be needed. For example seeds of *Quercus* are fumigated with serafume or other chemicals or heated in warm water for control of weevils, while methyl bromide or carbon bisulphide are also commonly used to kill insects.

Initial viability. Seed lots with high initial viability and germinative capacity have a higher longevity in storage than those with low initial viability. Germination tests, preceded if necessary by appropriate pretreatment to overcome dormancy, should be carried out on a sample of each seed lot before storage, in order to determine how long the seed is likely to retain viability in storage. Longevity of the viable seeds is correlated with the percentage which germinate in the initial test. As an example, samples of two seed lots of the same species, from which 80% germination of fresh seeds is normally expected, might give results of 90% and 50% initial germination.

Not only would storage of the second seed lot involve wasting space in storing dead seed, but even the 50% of initially viable seeds are likely to lose their viability more quickly in storage than would the 90% viable seeds in the first seed lot. Deterioration in initial viability may not be serious if the seeds are to be sown within a few weeks or months, but only good quality seed should be stored for long periods. In long-term storage of agricultural seeds for genetic conservation, it is recommended that no seeds should be accepted for storage which have an initial viability of less than 85% of that considered typical for the species or variety in question. It may be noted that initial viability and germinative capacity are frequently the resultant of the factors described in previous paragraphs (seed maturity, mechanical damage, fungal or insect attack).

Conditions of Storage and Ageing

In common with all other living things, seeds are subject

to ageing and, eventually, to death. In the case of orthodox seeds, the process of ageing and deterioration is so greatly affected by the conditions of storage that the "age" of seeds, expressed solely in terms of the period elapsed since ripening and harvesting, is an inadequate measure of the degree to which they have "aged" in the sense of losing viability and progressing towards the irreversible deterioration of death.

The term "physiological age" is commonly used to describe the degree of deterioration of seeds measured by their reduced capacity for germination. Nomographs for the effects of temperature and MC on physiological ageing of seeds have been constructed for several agricultural crops. As an example, the nomograph for barley indicates that the same degree of deterioration (from initial 95% to final 50% germination) would occur in about 16 days in seeds stored at 25°C and 21% MC, but in about 100 years in seeds stored at 8°C and 8% MC. Both seed lots would have the identical physiological age, though stored for very different periods of time.

Similar effects can be expected in orthodox seeds of forest trees. A number of physiological changes in cell tissues may be associated with physiological ageing in seeds. They include

(1) Loss of food reserves caused by respiration, e.g. decrease in proteins and non-reducing sugars, accompanied by increase in reducing sugars and free fatty acids

(2) Accumulation of toxic or growth-inhibiting by-products of respiration

(3) Loss of activity of enzyme systems

(4) Loss of ability of dried proteid molecules to recombine to form active protoplasmic molecules on subsequent rehydration

(5) Deterioration of semi-permeable cell membranes

(6) Lipid peroxidation, leading to production of free radicals which react with, and damage, other components in the cell

(7) Alterations to DNA in cell nucleus, causing genetic mutations as well as physiological damage.

It is still uncertain to what extent these various effects are the causes or only the symptoms of deterioration, but it has been suggested that the production of free radicals is the first effect of ageing and that damage to the several systems in the cells is the subsequent result of the release of free radicals. Whatever the exact mechanism of seed deterioration, there is a consensus that, in *orthodox* seeds, loss of seed viability is largely governed by the rate of respiration.

Any measures which reduce the rate of respiration without otherwise damaging the seed are likely to be effective in extending longevity during storage. These are the control of oxygen, the control of moisture content and the control of temperature. In *recalcitrant* seeds the safe minimum levels of oxygen, moisture content and temperature, and hence of respiration, are all considerably higher than those for orthodox seeds but, provided levels are maintained above the safe minima for each species, it appears that longevity can be extended by keeping them as close to the minima as possible in order to avoid an excessively high respiration rate.

Storage Atmosphere

The most obvious method of reducing the rate of aerobic respiration is to exclude oxygen from the atmosphere surrounding the seeds. This can be done by replacing oxygen by other gases such as CO_2 or nitrogen, or by using a partial or complete vacuum. In an example with lettuce cited by Roberts, seeds were stored in sealed containers at 6% MC and 18°C. After 3 years, seed stored in an atmosphere of pure oxygen had 8% viability, those in air 57%, those in nitrogen or argon or CO_2 78% and those in a vacuum 77%.

The value of excluding oxygen during storage of dry orthodox seeds has also been demonstrated in *Pinus radiata*. Best results were obtained with a storage atmosphere of nitrogen, followed by CO_2, while vacuum and air both gave poorer results. At the highest temperature used, 35°C, at which deterioration in viability was most rapid, the loss in final germination after 50 weeks' storage in sealed containers at 8% MC was 8% in nitrogen, 14% in CO_2, 21% in vacuum and 29% in air.

The same ranking was obtained by comparing the speed of germination and the vigour of germinated seedlings (measured as dry weight 49 days after sowing). Although increases in seed longevity of this magnitude have been achieved experimentally, some of the methods are expensive to apply and the effects on seed life are less dramatic than the effects of differences in temperature and humidity. Exclusion of oxygen will prevent aerobic, but not anaerobic, respiration, whereas reduced MC and temperature will decrease the level of both.

While systematic predictions have been made of seed longevity under a range of temperature and MC for several agricultural crops, similar quantitative predictions of the effect of oxygen levels on longevity are lacking. One simple method which is recommended is to fill sealed containers as nearly full as possible. If there is only a small amount of air inside the container as compared with the volume occupied by the seeds, oxygen will be consumed and CO_2 produced. The resulting high CO_2/O_2

ratio is probably favourable for seed longevity in orthodox seeds.

Whereas the complete exclusion of oxygen from the storage atmosphere appears beneficial to most dry orthodox seeds, there is evidence that some oxygen is necessary for recalcitrant seeds. Seeds of *Araucaria hunsteinii*, which had an initial germination of 56%, had all died within a month if stored in pure nitrogen, in two months if stored in 1% oxygen, and in three months if stored in 5% oxygen, while germination was still 18% after four months' storage in 10% oxygen. Storage in a polythene bag of 25 micron thickness, which was ventilated (21% oxygen) periodically when opened for extraction of samples, gave results very similar to those in 10% oxygen.

Moisture Content

The relationships of seed moisture content on a wet weight or fresh weight basis to seed MC on a dry weight basis, and of the equilibrium moisture content of seeds to the relative humidity of the surrounding atmosphere, are important in seed processing. They are equally important in seed storage. In the first case manipulation of RH can effectively *change* MC of seeds to the optimum for storage, in the second case MC can be *maintained* at or near that optimum by maintaining a suitable RH in the atmosphere around and between the seeds.

Effect of MC: In orthodox seeds, moisture content is probably the most important single factor in determining seed longevity. Reduction in MC causes a reduction in respiration and thus slows down ageing of the seed and prolongs viability. Harrington, cited by Barner, has related MC to various processes within and around the seed as follows:

Seed moisture content (wet weight)	%
Above 45 - 60%	Germination begins
Above 18 - 20%	The seed may heat (due to a rapid rate of respiration and energy release)
Above 12 - 14%	Fungus growth can occur
Below 8 - 9%	Insect activity much reduced
4 - 8%	Sealed storage is safe.

Prevention of fungal activity is more easily achieved by controlling MC than by controlling temperature. If MC and RH are high enough, fungal activity is possible between -8°C and +80°C and it is easier to keep MC below 12–14% (or RH to the equilibrium of around 65%) than to maintain sub-zero temperatures. Within the range of 4 to 14% MC, Harrington has suggested a rule of thumb applicable to many agricultural species—the life of the seed is doubled for every 1% decrease in MC, Schönborn found a relationship of a similar order when measuring the respiration rate, expressed in terms of CO_2 production, of *Picea abies*.

At 20°C and 20% MC the seed gave off 80 ml. CO_2 per hour per kg of seed; at 20°C and 5% MC, the rate of CO_2 production was reduced to 0.11 ml/hr/kg, a reduction of nearly a thousand times for a difference of 15 percentage points in MC. An MC of 4 - 8% is considered safe for most orthodox species; 5% ± 1% is recommended for long-term storage for genetic conservation. Oily seeds will usually tolerate drying to a somewhat lower moisture content (calculated on the basis of total fresh weight) than will non-oily seeds.

Drying below 4% can lead to damage or more rapid loss of viability in some species, although certain species can be dried to a considerably lower MC. *Betula papyrifera* was successfully stored at 0.6% MC without injury, while Schönborn succeeded in drying small samples of *Picea abies, Pinus sylvestris, Pseudotsuga menziesii* and *Larix decidua* down to 0% MC without any observable drop in germination after 6 months, compared with germination at the normally applied 6 - 8% MC.

Drying in this case was done, not by exposing the seeds to high temperatures but by leading a current of dry air through the seeds at 20°C. The same treatment killed *Pinus strobus* and *Abies alba,* and earlier attempts by Barton to store seeds of several species of *Pinus* and *Picea* at 0% MC also failed. Below about 2% MC desiccation injury becomes a strong possibility in many species. Drying to very low MC is also more costly than drying to the usual 4 - 8% and is likely to be used only in exceptional cases.

Some orthodox forest trees store best at appreciably higher MC. 8–10% is recommended for *Fagus sylvatica.* For seeds of *Abies* spp., an MC of 12–13% is recommended for storage of one to three years, but for longer periods this should be reduced to 7–9%. Species which benefit from storage at higher than average MC also need particular care in the timing and speed of drying.Fluctuation in the moisture content of seed in storage due to open storage without humidity control or to frequent opening and resealing of sealed containers results in deterioration in the germinability of the seeds.

In fact a steady MC slightly above the optimum is usually less harmful than one which fluctuates between the optimum and a higher moisture content. Some cases have been reported in which the usual trend of decreasing seed longevity associated with increasing MC is reversed at or near the moisture content of fully imbibed seeds. If the species in question needs exposure to light in order to germinate, it is possible to store fully imbibed but ungerminated seeds for some time in the dark. *Fraxinus americana* was stored at 22°C at varying MCs with the following results.

MC%	*Germination (%) after storage periods indicated*			
	1 month	*2 months*	*3 months*	*4 months*
6.0	98	92	96	94
9.5	94	88	76	4
18.6	81	22	0	0
Fully imbibed (in dark)	96	95	98	96

It has been postulated that imbibed seeds can repair damage to cell membranes, enzymes and DNA in the cell nucleus caused by free radicals in a way which is not possible for seeds at lower MC. Prolonged imbibed storage may, however, be difficult in practice, because of the need to maintain constant high moisture for imbibition and adequate oxygen without allowing the seeds to germinate or encouraging the multiplication of fungi and bacteria. Moisture content is also important in recalcitrant seeds, but in this case the critical MC is the minimum to which it is allowable to dry the seeds rather than the maximum content for prolonged storage.

For many of the large temperate hardwood seeds, moisture contents in the range of 25 - 79% are appropriate. Storage should be carried out at close to the minimum safe MC, since the higher the MC the higher the respiration rate and the more rapid the loss of viability. Higher respiration rates release higher amounts of energy and there is a risk of overheating and death of the seed, unless great care is taken to provide adequate aeration. High MC also increases fungal activity and the spread of rot.

Wang quotes results from two seed lots of *Acer saccharinum;* germinability of one lot stored at 58% MC and 1 - 2°C dropped from 94% to 12% after 6 months' storage, while that of the other, stored at 45% MC at the same temperature, was still 78% after 16 months. Loss of viability in this species may be sudden. Tylkowski found that seeds stored in sealed bottles at 50–52% MC and -1°to -3°C had over 90% germination after 18 months but this dropped to nearly zero after 24 months.

Less research has been done on tropical recalcitrant species, but there is some evidence e.g. in *Triplochiton,* that viability may be significantly prolonged if the minimum MC for the species can be determined and if particular care is taken over the period of drying down to that MC. Trials of *Shorea platyclados* in Malaysia indicated that a gradual reduction of MC to 20 - 27%, followed by sealing in charcoal, sawdust or vermiculite at 15°- 22°C allowed storage for at least one month, compared with the week or so of natural viability.

On the basis of experiments carried out on *Shorea parvifolia* and *Dipterocarpus humeratus,* Maury-Lechon et al. recommended reduction of MC to between one quarter and one half of the initial MC in freshly collected fruits. Although tropical recalcitrant seeds cannot yet be stored for more than short periods, there is a growing body of useful research on the problem. King and Roberts contains a good summary of achievements and possible approaches.

Storage Temperature

Temperature, like moisture content, is negatively

correlated with seed longevity; the lower the temperature the lower the rate of respiration and thus the longer the life-span of the seed in storage. Harrington suggested another rule of thumb for agricultural seeds—between 50°C and 0°C, every 5°C lowering of storage temperature doubles the life of the seed. For orthodox seeds, which can be dried to a low moisture content, still greater longevity can be assured by storage at sub-freezing temperatures.

For long-term storage for genetic conservation of agricultural seeds, a temperature of -18°C has been recommended as the "preferred" standard for most species and -10°C as an "acceptable" standard for those species known to have an intrinsic high viability. Much lower temperatures have been used with success on an experimental basis, e.g. in liquid helium at -269°C, but the high cost of maintaining such low temperatures for a long period would outweigh any (as yet unproven) advantages in increased longevity.

Choice of storage temperature varies considerably according to species and the period for which the seed is to be stored. The lower the temperature that has to be maintained in a cold store, the higher the cost, and provision of subfreezing temperatures may be unnecessary if seed is to be stored for only a year or two for afforestation projects. Holmes and Buszewicz noted that, in a number of experiments on conifers, the superiority of sub-freezing temperatures became evident only after prolonged storage over periods of about 5 years or more.

Sub-freezing temperatures (-18°C) appear to prolong viability in tropical orthodox species such as *Eucalyptus deglupta* and *Flindersia brayleyana*. Some seeds keep well at room temperature, e.g. many leguminous and rosaceous genera, *Eucalyptus, Tilia* and many other hard-seeded or stone fruits. Most species, however, only keep well for longer periods at lower temperatures. In 3 - 5 years' storage of most conifers and *Alnus* and *Betula, the crucial temperature seems to lie at a maximum of* +4°C.

The temperature should therefore be kept at 1 to 4°C. For longer periods of storage, say 5 - 15 years, the temperature should be -4 to -10°C. For *Abies,* however, a temperature of -4°C is used for short storage periods and -10°to -20°C for longer periods. Temperature and moisture factors are so interrelated that it is very difficult to separate them. Seeds at a relatively high moisture level can be stored for considerably longer periods at near freezing temperatures than at higher temperatures, while higher storage temperatures (30°C) are less harmful when the moisture content of the seed is low.

In short, it can be said that the critical moisture content lies at a higher level when storage temperatures are low than when they are intermediate or high, i.e. to some extent, *a low temperature can compensate for a high moisture content, and vice versa.* It is, however, necessary to avoid any risk of freezing damage caused by ice formation in seeds of high MC. The equilibrium moisture content of many seeds at a given RH varies with temperature. At a range of RH from 35% to 76%, the amount of water contained by seeds increased progressively as temperature decreased from 30°C to 10°C.

Species included *Pinus* as well as several agricultural crops. At low (35%) RH moisture absorption by dry seeds was approximately the same at 5°C as at 10°C, but at higher (55% and 76%) RH the seeds absorbed less moisture at 5°C than at 10°C, thus reversing the trend above 10°C. Change of equilibrium moisture content with change of temperature can be of importance in open storage. In sealed containers the effect is minimal because the final EMC is dominated by the initial MC of the seeds and not the moisture of the enclosed air.

As with moisture content, repeated fluctuations in temperature lead to loss of viability. As far as possible, temperature should be maintained at a uniform level. The effect of temperature on seed longevity of temperate recalcitrant species is similar to that of orthodox species—within certain limits the lower the temperature the longer the period of viability. Some tropical species are killed by temperatures above freezing e.g. some dipterocarps at <14°C, cacao at <10°C and mango at < 3–6°C. Seeds of *Hopea helferi* stored at 15°C with high moisture content in unsealed polythene bags retained 98% germination after 37 days and 80% after 60 days.

Germination was much reduced if temperature was dropped to 10°C or raised to 25°–28°C. *Shorea ovalis* is another species which does not withstand low temperatures; it stores best at 21°C. *Shorea talura*, in contrast, stores well at 4°C and 40% MC (wet weight basis); after six months germination was reduced from an initial 95% to 69%. Other species of dipterocarp have a much shorter seed life. For most temperate recalcitrant species the lower the temperature down to 0°C the longer the safe storage period.

Below-freezing temperatures, on the other hand, often kill recalcitrant seeds which need to be stored at a high moisture content. Some success has been achieved in storing *Quercus* seeds in the USA by maintaining them at 35 - 45% moisture content and -1°to +3°C temperature. Temperature can be critical because less than -1°C will usually kill the seeds, while temperatures above 2°or 3°C cause excessive germination. In Europe more northerly species of *Quercus* can be stored at slightly lower temperature -1°to -3°C and 38 - 45% MC.

Light

Light, particularly ultra-violet light, is reported to be harmful to seed, but very few studies have been made. Use of opaque metal containers would be preferable to glass jars or bottles for species which are affected by light. But light appears to be much less important than either moisture content or temperature.

Method for Storage

A number of different storage methods are available, as described below. The main factors affecting choice are the seed characteristics of the species in question, the period for which it is to be stored and the cost. If more than one method is suitable to maintain viability for the required period, the simplest and cheapest will normally be chosen.

Storage at Ambient Temperature and Humidity

Seeds can be stored in piles, single layers, sacks or open containers, under shelter against rain, well ventilated and protected against rodents. Best results are obtained in cool, dry climates. In these conditions several species of *Pinus, Eucalyptus Pseudotsuga* and *Tectona* will store satisfactorily for at least six months, while leguminous trees with impermeable seedcoats and naturally low MC, e.g. *Acacia Prosopis, Robinia* will retain viability for years.

Dry Storage

Orthodox seeds will retain viability longer, when dried to a low moisture content (4 - 8%) and then stored in a sealed container or in a room in which humidity is controlled, than when stored in equilibrium with ambient air humidity. Storage life is further prolonged if cool, but not controlled, temperature conditions can be provided, e.g. at high latitude or altitude and in a cellar or other room screened from direct sunlight. Seed is sometimes stored in open containers in a room maintained at an RH of 15 - 20% by dehumidifying machinery.

In forestry it is more common to rely on predrying of the seed to the correct MC, followed by storage in full, sealed containers. Provided that the containers are not opened too frequently and that the sealing is effective, the method will maintain a low MC for many years. It is cheaper than using a humidity-controlled room, especially during periods when only a little seed is in store, and it is not subject to hazards of mechanical breakdown. This method is suitable for a range of species including many *Pinus* and *Eucalyptus* species, for which it should maintain viability for one or more years.

Dry Storage with Control of both MC and Temperature

This storage treatment is the standard practice for many orthodox species which have periodicity of seeding but which are planted annually in large-scale afforestation projects. For many species a combination of 4 - 8% MC and 0 to +5°C temperature will maintain viability for 5 years or more. Some cool-temperate genera benefit by storage at sub-freezing temperatures, e.g. -4°C or lower for *Abies*, -10°C for *Fagus*, -5°C for *Fagus*, -18°C for *Pinus strobus, Populus deltoides* and others.

Pinus merkusii is an example of a tropical pine which responds well to storage at low temperature and MC. 80% germination was obtained after 3 years' storage at 2°C and 6 - 10% MC of seeds of the Zambales (Philippines) provenance, while seeds stored in equilibrium with room temperature and humidity showed a significant loss of germination after 3 - 4 months. *P. caribaea* and *P. oocarpa* behave similarly in this respect. In addition to true seeds, this method is also suitable for certain types of fruits. For example in the Jari project of Brazil depulped, cleaned and dried stones of *Gmelina arborea* are successfully stored in sealed containers at 5°C and 6 - 10% MC. The fresh stones have a germination of 90% and after two years of storage germination is still 80%.

The preferred storage treatment for long-term conservation of gene resources of orthodox agricultural seeds is -18°C temperature and 5% +-1% MC. This is likely to be equally appropriate for orthodox seeds of forest trees requiring storage for genetic conservation. The quantity of seeds requiring this standard of storage is small in comparison with the quantities used each year for operational afforestation, and the cost per kg of seed is higher. For many countries it would therefore be desirable that forest and agricultural crop genetic resources should share a common long-term storage facility.

A good example is that of the Banco Latino Americano de Semillas Forestales at CATIE, Turrialba which has its own seed store (55 m^3 capacity at 5°C) for short or medium term storage, but also has access to the long-term storage facilities (at -20°C) of the Regional Genetic Resources Unit, which are also at Turrialba. Loss of viability in storage, in addition to reducing the number of plants which can be produced by a given seed lot, may result in a shift in the genetic constitution of the seed being stored. This could be particularly important in forest trees which are predominantly outbreeding, variable populations.

First, loss of viability may occur more rapidly in some genotypes than others; if losses are high, say 50% of the total, the genotypes with short-lived seeds may be eliminated altogether. Yet they may have valuable traits for adaptation, growth or disease resistance as growing trees, and in any case they contribute to genetic variation in the species which it is the purpose of genetic conservation to preserve. Secondly, it is an accepted fact in agriculture that chromosome damage or change occurs and accumulates in the seed in storage, and that the risk of such heritable gene mutations depends not so much on the age of the seed as on changes in its viability.

A seed lot which has suffered a serious loss of viability is likely to have experienced some gene mutations among the survivors, but there is very little direct evidence that heritable mutations are induced under good storage conditions which lead to only small losses in viability. The high standard of storage conditions recommended by the IBPGR and referred to above, if combined with regular testing of seed and regeneration as soon as germination falls to 85% of the initial germination rate should minimise the risk of genetic change in storage.

It is possible that still lower temperatures would increase longevity even more. Research on storage in liquid nitrogen has been pursued for some years and considerable progress has been made, but testing for several more years will be needed before the method can be recommended for general adoption in gene banks.

Suitable for storage of recalcitrant seeds for a few months over winter. Seeds may be stored in heaps on the ground, in shallow pits in well-drained soils or in layers in well ventilated sheds, often covered or mixed with leaves, moist sand, peat or other porous materials. Seeds stored outdoors are kept moist by rain or snow, but those under shelter may need to be moistened periodically.

The aim is to maintain moist and cool conditions, combined with good aeration to avoid overheating which may result from the relatively high rates of respiration associated with moist storage. This may be accomplished by regular turning of the heaps of seed or by inserting bundles of straw or twigs into them. This method is suitable for short-term storage of large-seeded hardwood species in the temperate zone e.g. *Quercus, Castanea, Aesculus*. It is unlikely to be suitable for tropical recalcitrant species because ambient temperature is too high.

Moist Cold Storage, with Control of Temperature

This method implies controlled low temperatures just above freezing or, less commonly, just below freezing. Moisture can be controlled within approximate limits by adding moist media e.g. sand, peat or a mixture of both to the seed, in the proportions of one part media to 1 part seed by volume, and remoistening periodically, or more accurately (but more rarely) by controlling the relative humidity in the cold store. The latter type of control is often too expensive. Respiration rate is reduced and storage life prolonged by the low temperature, but seed should not be stored in sealed gas-proof containers which would limit oxygen supply.

Closed polyethylene bags of 4 - 10 mil (100–250 microns) thickness will allow exchange of oxygen and CO_2 with air outside, while severely restricting exchange of moisture. Sub-freezing temperatures have given improved results in a few cases but frequently injure seeds with high MC and should be used only after research has demonstrated their applicability to the species in question.

Less is known about the application of this method to tropical species, but it merits much more investigation that it has so far received, for the dipterocarps and genera such as *Araucaria, Agathis* and *Triplochiton*. As mentioned earlier there is evidence that some species are killed at low but above-freezing temperatures and Gordon has proposed a division of recalcitrant seeds into those which can withstand temperatures below 10°C without loss of viability and those which cannot.

Tamari, summing up several years of research on dipterocarps in Malaysia, concluded that the best treatment for several species was (1) Dry at a temperature not above 35°C, to reduce MC to 35% (2) Seal with a fungicide inside polythene bags (3) Store at 15°C, or for 3 weeks at 15°C followed by further storage at 10°C. This treatment has been successful in extending longevity from a week or two up to two months in *Hopea helferi*, but this is still a long way from providing safe storage between seed years, reported to vary from 3 - 6 years apart in many dipterocarps.

Storage at 3.5°C and MC over 32% over periods of at least 6 months has been successful on the recalcitrant *Araucaria hunsteinii*. In some recalcitrant species newly germinated seeds may retain viability better under moist cool storage conditions than ungerminated seeds. Gordon reported that some pregerminated seedlots of *Quercus* spp. showed no significant change in the number of living seedlings after one year's storage in 500 gauge (125 microns) polythene bags lightly sealed at 3°C, whereas a large proportion of the seeds which were viable but ungerminated when placed in the same bags died during the same period.

Other Storage Methods

Other storage methods have been used in the past, but are not yet of wide application. They include:

a) Storage of recalcitrant seeds in running (not stagnant) water.

b) Storage under partial vacuum.

c) Storage in gases other than air e.g. nitrogen or CO_2.

d) Coating individual large seeds with paraffin or latex to prevent moisture exchange. This method may also be used to maintain moisture content during shipment.

Storage Containers

Some form of container is necessary for most seed storage, to facilitate access to, and handling of, individual seed lots while keeping them separate, to make the best possible use of storage space, to provide protection against animal and insect pests and, for some seeds, to prevent passage of moisture and gases between the enclosed and the outside atmosphere. Many types of container have been used for tree seeds; they may be conveniently divided into (1) Materials freely permeable to moisture and gases (2) Materials completely impermeable, when sealed, to moisture and gases (3) Materials resistant, but not completely impermeable, to moisture.

Materials Freely Permeable to Moisture and Gases

These include hessian or burlap sacks, cotton bags and containers of paper, cardboard and fibreboard. Hessian and cotton have the advantage that seed triers can be inserted through the cloth mesh to withdraw samples for testing without the need to open the mouth of the container. The resilience of the cloth will close the hole and avoid subsequent loss of seed, which is not possible with containers based on paper or paperboard. Hessian and cotton are also robust materials which can be used more than once. None of these materials is entirely proof against insect and rodent pests, and all are freely permeable to water vapour and gases.

For orthodox seeds in uncontrolled conditions they are therefore suitable only for rather short storage periods; these can be extended in the case of hardcoated seeds or where ambient conditions are cool and dry. If seeds are stored in large containers after drying to the correct MC, the outer seeds themselves provide some barrier to the passage of moisture. Viability of the inner seeds may thus be preserved for a period even though there is some deterioration from increased MC in the outer layers.

If a seed store has facilities for controlling both temperature and relative humidity, then permeable containers can be safely used for orthodox seeds for several years, provided that pests can be excluded. For moist storage of recalcitrant seeds, open or freely permeable containers such as hessian sacks should be used in order to allow free exchange of air and so avoid the overheating which can occur if moist, rapidly respiring seeds are enclosed without adequate ventilation. Periodic spraying of the sacks may be necessary to maintain the high MC which is appropriate for this type of seeds.

Materials Completely Impermeable, when Sealed, to Moisture and Gases

After drying of orthodox seeds to the correct MC, the MC may be maintained in storage by dehumidifying the whole storage space. Another very efficient way, commonly used in storing forest seeds, is to place the seed in sealed moistureproof containers. This avoids the need for expensive dehumidification equipment. For long-term storage the most effective method is a combination of moistureproof containers with controlled low temperatures provided by refrigeration.

An added advantage of most materials in this type is that they also exclude oxygen and so reduce still further the rate of respiration. Impermeable sealed containers are not suitable for storing recalcitrant seeds nor are they suitable for orthodox seeds at high MC, which deteriorate more rapidly in sealed than in open storage. Some seeds absorb moisture quickly, so it is important that they be sealed inside the container as soon as possible after drying is complete, preferably within the drying room itself.

Moistureproof containers include tin or aluminium cans and drums, glass jars of the Mason or Kilner types, plastic vials and laminated aluminium foil packages. Rigid and unbreakable metal cans provide maximum protection against mechanical damage to the seeds and are equally suitable for storage and subsequent shipment. Containers are only as moistureproof as their sealing. For rigid containers screw-top or clamp-down gasketed lids should be used, if periodic opening for seed extraction and subsequent resealing are anticipated; aluminium foil should be heat-sealed.

The effectiveness of sealing is particularly important in long-term storage. Three types of container are considered suitable for hermetically sealed long-term storage of agricultural seeds: glass jars or vials; metal cans; and laminated foil packets. They should be equally appropriate for orthodox forest seeds. But the report by IBPGR recommended sealed metal cans as the most reliable and convenient. It noted that the seals on screw-cap jars are not always perfect and that further experience of the lasting qualities of laminated foil packets is needed before they can be recommended for general use in storage which will often last for several decades.

Materials Resistant, but not Completely Impermeable, to Moisture

They include polyethylene and other plastic films and aluminium foil. These materials are resistant to the passage of moisture but, over a long period of time, there will be a slow passage of water vapour tending to equilibrate the RH inside with that outside the container.

The RH gradient during storage is never so severe as this and hence the rate of passage of water vapour is much less rapid in practice. In one test using 6 mil (150 microns) high density polyethylene the rate of passage over two years from an outside RH of 95–100% at 20°/30°C was four times that from an outside RH of 50% at 10°C. The thicker the film, the greater the resistance to passage of water vapour and, for a given thickness, high density polyethylene is more resistant than low density.

Although polyethylene is not suitable for long-term storage of orthodox seeds for genetic conservation, it is very suitable for short-or medium-term storage and has given excellent results for up to 5 years' storage of *Pinus caribaea* and *P. oocarpa* seeds in Honduras, with no significant change in MC. For Honduran conditions a thickness of at least 4–5 mil (100–125 microns) is recommended; thinner polythene can permit a significant passage of water vapour in time and is also subject to mechanical damage in handling.

Use of Desiccants in Containers

If orthodox seeds are dried to the correct MC and stored in sealed impermeable containers, the MC should remain constant for years. If, however, the seeds are stored in moisture resistant but not completely impermeable material such as polythene bags, or if it is necessary to open and reclose the containers periodically to extract seeds, there will be a slow build-up of moisture in time. A convenient way to prevent this is to enclose some desiccant such as silica gel in the containers. The capacity of silica gel to adsorb moisture depends on the relative humidity of the ambient air, as shown in the following table:

Moisture content of silica gel in equilibrium with various relative humidities

% RH	*% H_2O adsorbed*	*% RH*	*% H_2O adsorbed*
0	0.0	55	31.5
5	2.5	60	33.0
10	5.0	65	34.0
15	7.5	70	35.0
20	10.0	75	36.0
25	12.5	80	37.0
30	15.0	85	38.0
35	18.0	90	39.0
40	22.0	95	39.5
45	26.0	100	40.0
50	29.0		

A convenient method is to use silica gel treated with cobalt chloride, which changes colour from blue to pink about 45% RH; the corresponding equilibrium MC for many orthodox species would be 7–9%. Dried silica gel is enclosed with the seeds and, whenever the granules turn pink, the silica gel is removed and reactivated by drying in an oven at 175°C and cooling in a sealed container before reuse. A weight of silica gel equal to one tenth the weight of seeds is recommended. Care should be taken not to include too much silica gel which could lead to overdrying of the seeds.

Even with silica gel of one tenth the weight of seeds, the MC of seeds enclosed at 6% would be lowered to below 5% during the initial phase of storage. More frequent reactivation of silica gel would preserve an equilibrium of RH and seed MC at lower levels than the 45% and 7–9% mentioned above but would forego the convenience of the colour indicator. Another use for desiccants is where the MC of seeds is known to be higher than the optimum for sealed storage, for example because only air-drying is possible.

Selection of Container

The following factors, which should be considered when choosing the best storage container for a given use, are based on those listed by Stein et al.: When seed requires further drying in storage, do not use a tight-closing container because enclosing excess moisture is harmful to the seed. Use a tight-closing container if gain in seed moisture content can be damaging and relative humidity in the storage facility is high.

Containers and seed can quickly gather unwanted condensation when brought out of cool or subfreezing storage. Warming to room temperature is recommended before opening a container brought out of such storage. 4 to 10 mil (100 - 250 microns) polyethylene bags will greatly restrict exchange of moisture, but still allow exchange of oxygen and carbon dioxide with air outside. Such exchange may be beneficial or harmful, depending on species.

A container that is easy to open and close is desirable when quantities of seed are likely to be added or removed repeatedly. In order to minimise temperature and relative humidity fluctuations, open only when necessary. Alternatively, store seed in small containers, so that the entire contents can be stored or emptied at one time. For orthodox seeds, fill containers completely to ensure minimum exchange of moisture between the seed and the entrapped air and, more importantly, to limit the amount of oxygen enclosed.

When exchange of moisture through the container walls must be eliminated or restricted, the container must be made impermeable or of moisture resistant material. The longer the storage period and the higher the differential between external RH and RH within the container, the more impermeable the material must be. When seeds are fragile and easily damaged, a rigid-walled container should be used. Moisture-proof plastic bags are often used as liners for rigid containers.

Choose a container shape and stacking arrangement which facilitates uniform temperature and aeration throughout the storage facility. Some containers may be made of substances that are harmful to tree and shrub seeds. Unproven containers should be tested for toxicity. Static electricity can build up slowly on some materials such as PVC. This makes them difficult to clean between exhaustion of one seed batch and insertion of the next. It should be stressed that no one container or packaging material can be the best for all sizes, conditions and objectives that occur in seed packaging. The relative merits of the various containers will have to be weighed against their disadvantages and costs before deciding the final choice.

8

Forest Seed Handling

Seed Production

In recent years programs of afforestation and reforestation have been planned and carried out at an increasing rate in many areas of all continents. This has resulted in a remarkable increase in the demand for forest tree seeds, which are often quite difficult to find, prepare and handle.

When artificial regeneration was first applied in forestry, the only problem was that of getting seed or nursery stock of desirable forest species, regardless of its genetic worth. But substantial failures of many large plantations, as well as several trials and investigations, proved the basic importance of the seed origin, which is today recognised everywhere by forest geneticists and research workers. Some forest administrations have prepared special laws in order to insure effective control of the forest seed provenance; but even nowadays many countries do not attach sufficient importance to this question.

The first step to insure provenance control is to identify, for each important tree species, a number of production areas or stands which, according to certain criteria, may be considered as suitable seed sources. Seed stands should consist mostly of very good trees, preferably middle-aged (generally 35 to 40 years is a minimum for conifers). Good stem form, small branches, rapid growth rate, and soundness are the main characteristics required. Moreover, it should be kept in mind that seed production is influenced by stand density, and that the stand should cover an area large enough to make collection convenient from an economic point of view.

Seed could also be collected from individual trees which combine to the highest degree the desired characteristics ("plus trees" or "elite trees"). But the collection from individual trees is usually carried out only for scientific purposes. Lately, the development of forest genetics and tree breeding has led to the concept and establishment of seed orchards. Seed orchards are plantations of genetically superior trees, isolated to reduce or prevent pollination from outside sources. They are established by setting out clones (as grafts or cuttings) or seedling progeny of trees selected for desired characteristics.

When seed orchards are established while genetic evaluation of the parent material is still under way, components shown by progeny tests to be undesirable should be removed from the orchards. Seed orchards are expensive to establish, but they can yield a large quantity of superior seed some years after their establishment. Seed collection is less expensive and the use of special harvesting equipment is possible.

Seed-bearing of forest trees is usually rather irregular from year to year. Only a very few species bear good seed crops every year. Generally one year of a heavy crop (seed year) is followed by one or several years of poor seed crop or none at all. Many conifers show intervals between seed years of three to four years. Most spruces, for instance, bear good crops at intervals of five years and occasionally at longer intervals up to seven years, light crops being rather common in the intervening years.

In order to organise seed collection properly it is therefore very important to forecast fruit-setting some months before collecting time. The estimates of seed crops are generally based on counting or weighing cones or fruits on a number of sample trees. The results are applied to the stand and are finally expressed as numerical values, going from 1 (total crop failure) to 10 (maximum crop, exceptional seed year). The ripeness of seeds and the beginning of natural seed dispersal are the main factors which limit the time suitable for collection.

Seed Ripeness

Seed ripeness is often judged by the colour of fruits or cones, but a better indication is provided by their specific gravity. The specific gravity of cones or fruits decreases as the seed approaches maturity. A mixture of various liquids may be adjusted so that ripe cones of a given species Will just float. Cones may also be tested in a set of liquids, with different known specific gravities.

The surest check of seed ripeness is to cut cones or fruits in two lengthwise, and examine the seeds which occur on the cut surfaces. By this method the number of fully developed seeds per fruit or cone can be counted at the same time. A good knife is often sufficient for this purpose, but special knives have been prepared which prove very useful for most cones and for some tougher fruits of hardwoods.

Seed Collection Methods

Collection Time

Seeds or fruits should usually be collected just as they have reached full ripeness, and before natural dispersal begins. This interval varies considerably in the different tree species, and with local climatic conditions. For many conifers, especially spruces, long autumn rains or fogs delay cone opening for a long time, while a few dry windy days are sufficient to give full seed dispersal.

Collection on the Ground

Seeds and fruits can be picked up from the ground after shedding. This is the case of many large-sized seeds of broadleaved trees, such as oak, chestnut, and beech. If possible, the ground should be cleaned from brush, and a, sheet placed under the tree makes collecting much easier. Seeds should be picked up as soon as possible after they have fallen to avoid losses and damage by insects, rodents or fungi; but it should be kept in mind that the first seeds or fruits which fall naturally are often of poor quality.

For all trees, a very easy and cheap method of collection is to take account of normal fellings occurring during the seed ripening season, and to collect seeds or fruits from felled trees. This method is largely used in northern Europe. In North America seeds are sometimes collected in the hoards of small rodents, which often gather large quantities of different seeds, especially nuts.

Collection from Standing Trees

The greatest quantity of tree seeds is collected from standing trees. This method is clearly often difficult and expensive, but it ensures that seeds come from selected trees and from the best part of their crown. In the case of shrubby or very small trees, seeds can be picked directly from the branches by the collector while standing on the ground. For small or medium-sized trees, poles, pole saws or trimmers are used for collection, the operator standing on the ground or on the lower branches of the tree.

Hooks of different shapes, cutting edges or little saws are attached to a long pole. By moving the pole, fruits or cones can either be detached from the branches, or the twig (in the case of small fruits) can be cut down completely. Sometimes a sack or a basket with a wire end can be used for fruits which are easy to detach. The length of the pole (that is the height which can be reached) depends on the weight and rigidity of the pole itself. Different materials are used:

a) bamboo is very light but not sufficiently rigid;

b) steel pipes are too heavy;

c) aluminium or plastic poles often give good results.

For large-sized trees, the normal method of collecting seeds is to climb the tree. Some forest workers are very good tree climbers, and climb tall trees without, any equipment. But this method, besides being dangerous, tends to make the operator choose trees for their climbability and not from the point of view of seed quality. If possible, good seed trees should be marked by a skilled technician, and workers should not be allowed to climb trees other than the marked ones. Low branched trees are easily climbed, and a normal climber is able to climb a trunk, even without branches, for a few meters, if the tree diameter is not too big (not over 40 centimeters). Otherwise, some equipment, such as irons or ladders, is required to reach the lowest branches.

Climbing equipment

Foot-clamps, loot-hooks or climbers. These are some of the usual types of climbing equipment: a forged iron is fastened by a belt to the climber's shoe, and finishes in a pointed spur, the length varying according to the climbing method. One of the best types is a German one, where the hook is quite short and does not extend beyond the shoe sole, permitting the climber to walk on the soil without difficulty. The irons always cause some damage, especially to very thin-barked trees: round points seem to be preferable from this point of view. Irons should not be used when the bark is frozen.

Ladders: Many types of ladder are used in tree climbing. Type, form, length and material vary largely according to tree species and local conditions. In level country and where forests have an extensive road network special ladders, which are sometimes heavy, can be transported by truck. In mountain forests transport difficulties arise and only small, light ladders may be employed.

For smaller trees, a light wooden ladder is perfectly adequate; types used by fire services, 6 to 8 meters long, do not weigh more than 2 kilograms, and can also be transported in mountain regions: they have to be used in a vertical position, i.e., parallel to the trunk. Naval rope ladders are sometimes used: a finer cord is thrown to a strong branch, and the rope ladder or cord is then fixed to the branch.

Greater heights require scaling ladders which can be assembled and dismantled. They may be of different types and materials: wood, aluminium, steel, and magnesium alloy are used. The elements of the ladder have a variable length (2 to 4 meters) according to their weight, which should not exceed 3 to 4 kilograms per element, to ensure easy transport during climbing. A small, adjustable platform is placed under the base of each leg, to ensure stability on the ground.

Each element is fitted easily and safely into the lower one and, when in place, it is fastened to the trunk by a chain or a belt which the operator fixes while ascending. Assembled scaling ladders may have one or two legs: two-legged ladders are more common and more easily purchased or constructed. Elements may be connected without limit: this type has been used up to a height of 38 meters. However, the rungs could be more widely spaced, and the connection improved. One-legged ladders are useful on steep or rocky ground where it is difficult to lodge a two-legged ladder evenly, and also in the case of a tree with many branches where it is hard to place a two-legged ladder along the trunk. Small bars are fixed as rungs alternately on either side of the central support. The form of the rungs varies with the type of ladder.

One-legged ladders can be attached by chains very close to the trunk against which they are held by a special spacer. These ladders can be attached to trunks which are not quite straight and continued up into the crown. They are easier to place on the ground than two-legged ladders, although transport during climbing is not so easy; but no special difficulties arise. The weight and length of the elements do not differ much from those of two-legged ladders, because the central support is generally in steel tubing. One-legged ladders give very good service, and are widely used especially in Germany.

Extension telescopic ladders may be placed and transported by special trucks. Fire-service extensible ladders are used to enable the worker to reach heights of from 20 to 25 meters. They may only be used in forests in the plains or for collecting seeds on trees growing at roadsides. An extension ladder in light alloy, which may reach a length of 14 meters, is mounted on a tractor, and can therefore be used also in forests without roads. A special model has been developed by the United States Forest Service. The truck is 0.5 ton pick-up on which a rectangular framework is mounted, supporting a heavy-duty aluminium extension ladder, 12 to 15 meters long.

Telescopic platforms rolling on wheels sliding under carriage or mounted on special trucks are also used. A special apparatus for tree climbing is the Swiss Baumvelo". It is a device for climbing straight branch-free trees, without damaging them, and is preferable to a ladder because it is much easier to transport and to use in dense forest stands. It consists of two similar parts, one for each leg.

A vertical arm supports a pedal in which the operator's shoe is fastened: at its upper end it is connected to a steel belt, forming a circle of adjustable diameter around the stem; a supporting block is fixed in the middle of the arm. After reaching the crown, the climber can easily leave the Baumvelo, attached to a branch, and climb freely into the crown, special safety belts being provided. This device has been developed by the Eidg. Forstliche Versuchsanstalt, Zurich, and has been favourably received in various countries, especially in the United Kingdom, where the Forestry Commission has improved the related safety equipment.

Another tree climbing device is the tree net, developed in the United Kingdom by the Forestry Commission. A triangular net is suspended by special ropes and snatchblocks from the top of the tree, and covers a large part of the crown; the collector can thus easily reach all parts of the crown. This equipment seems to give good results for large branched trees bearing small fruits which take a long time to collect. Special equipment is available for rigging the net, but the operation is time-consuming and is very difficult, or quite impossible, in dense stands. Sometimes permanent or semipermanent scaffolding, in iron pipes or wooden poles, is built around the trees; this is done mainly for scientific purposes, to collect seeds or scions or to control the pollination of plus trees.

Seed Transport

Collected cones or fruits are put into sacks generally with a capacity of 1 hectoliter (3 bushels), corresponding to about 50 kilograms (100 pounds) in weight. Sacks or other containers should be filled and closed under proper supervision, and their number or weight checked, to ensure provenance control. If cones or fruits cannot be transported at once to the storage point or factory, provisional storage should be arranged locally, in huts or under some kind of shelter; in any case sacks or containers should not be stored in large amounts over a long period.

Various kinds of transport can be employed. Trucks are generally used, but railway transportation is more economical over very long distances. While loading trucks or cars, care should be taken to ensure proper air circulation among sacks. Over short distances, cones could be loaded onto the truck without containers: in this case, each truck should carry cones or fruits of one single provenance and species.

Seed Extraction

The word "extraction," in the wider sense, includes all the different operations or processes necessary after collection to prepare seed for sowing. This preparation usually consists in separating the seed from all or some parts of the fruit. It is easy to understand that extraction technique varies greatly according to the species: some seeds are quite ready for sowing as collected from the tree; others are obtained after a long series of difficult processes, i.e., opening of cones, separating seed from wings, etc.

Extraction usually consists mainly of one or more of the following processes: washing, maceration, threshing, drying, separation, de-winging, cleaning. Some of these operations could be carried out in a very simple manner, and do not require any special equipment; but often, and in any case when large amounts of seeds are to be prepared, a special extractory must be established, with stores, machines and any other equipment required for the various operations.

Hitherto seed separation has been carried out by very simple means in most countries, so that machines specially made for forest seed handling are difficult to obtain, and often agricultural machines have had to be adapted for this purpose. During recent years, however, some special machinery has been developed and constructed.

All operations, whether for extraction of small or large amounts of seed, must follow a certain sequence. It is clear that certain processes could sometimes be delayed, or left out altogether in the case of occasional operations, while in the case of a permanent seed plant with large output a careful organisation of the working cycle is absolutely necessary. In any case extraction should follow a general line, which is briefly described below.

Cones or fruits coming from collecting points go to pre-cleaning, and pass at once, or after storage, to extraction. Fleshy fruits are submitted to maceration, cones are dried (naturally or artificially) and tumbled, other types of fruits are threshed. After extraction, seeds are de-winged (if necessary), cleaned and put in storage. A seed plant should therefore include the following rooms and equipment:

1. reception room, and precleaning room and equipment (preferably on the ground floor)
2. stores for fruits or cones
3. a local store (placed above the. kilns or extractors) supplying machines
4. kilns or yards to dry cones or fruits, respectively artificially or naturally
5. extraction room and machines (including de wingers and cleaners)
6. seed stores.

Rooms and equipment should be placed so that fruits and seeds pass from one operation to the next in the easiest and safest way, providing as far as possible mechanical transport between machines. A two-storey building is often recommended. The working material comes to the upper floor from the precleaning plant or from the cone or fruit stores. It is placed here, in a local store, directly supplying kilns which are placed on the first floor. From the kilns seed falls down to the ground floor where separation and cleaning equipment is located.

Stores for fruits or cones and for seed are generally placed in one or more separate buildings. The size of any equipment, installation or plant should be carefully studied and planned in relation to the amount of seed required per year. It should be borne in mind that many species do not give a good seed crop every year, so that in good seed years it is necessary to collect and process a supply of seed for two or three years. Therefore, the equipment must be proportionate to the amount of fruits or cones collected in such years.

Even if good storage facilities for fruit or cones are available, it is preferable to keep the working season as short as possible, to avoid storage difficulties, deterioration, and loss of seed quality. The interval between fruit collection and seed preparation is strictly related to the local climatic conditions. The choice of machinery should be made after considering the output per hour required, which may be calculated by the formula:

$$W = \frac{S \times F}{Y \times P \times T}$$

where:

W = is the working capacity of the machinery (in hectoliters of cones per hour;

S = amount of seed required each year (in kilograms)

F = average frequency of seed crops (in years)

Y = seed yield (in kilograms) per hectoliters of cones or fruits

P = working period (in days)

T = daily working time (in hours).

On the basis of this formula and given the following data, the resulting output per hour would be 1 hectoliter.

Amount of seed required each year	= 1,000	kilograms
Average frequency of crop	= 3	years
Total amount of seed to be prepared in seed years	= 3,000	kilograms

Seed yield per hectoliter of cones or fruits	= 2.5	kilograms
Corresponding amount of cones or fruits to be processed	= 1,200	hectoliters
Working period	= 100	days
Output per day	= 12	hectoliters
Daily working time	= 12	hours

For small amounts, it is often more economical to prepare seed by hand or with very simple equipment. But for very large seed production, all modern machinery and auxiliary equipment will greatly reduce costs: mechanical elevators, transport bands or blowing pipes give positive results and often allow a better handling of the seed. Sometimes, a choice must be made between a certain number of small plants placed in the forest and dealing with a limited amount of seed from the local provenances or a large central plant, handling seeds from a wider area.

Such small plants, where work is usually done by hand, could be equipped with relatively simple machines which reduce time and transport costs; they are preferable for species requiring quite simple handling. Large central plants are run more economically if very large amounts of seed are to be produced, especially for species which require a long series of difficult processes.

In a large plant it is possible to establish all types of equipment, even if very expensive. Full mechanisation and automatic control ensure good organisation of the work and eventually a continues working cycle with an increase in the daily output of the plant. Of course, such plants demand a large capital investment, and costs should be carefully studied. It is probable that they are profitable only if the yearly production can be estimated at several thousand kilograms of seed.

Precleaning and Storage of Fruits

Cones or fruits, after transportation from the collecting place to the working plant, should be cleaned before going to the kilns or extractors, or to the stores, since they are often mixed with pieces of bark, wood, and twigs or other impurities. This rough cleaning may be done by flat or by rotating screens. Another method of precleaning is by flotation: damaged fruits and impurities come to the water surface and can easily be skimmed off. But the commonest method is the use of oscillating screens or vibrators.

For seeds of some trees, as *Carya, Juglans, Quercus, Ulmus,* cleaning is the only handling required. Storage of fruits or cones should be planned with care. Many fruits, especially conifer cones, need a period of storage to become completely ripe. If a large amount of any fruit is collected in a very short time (for some species the collecting season is not more than two weeks since seeds fall quite early when ripe) it may be necessary to store it for some months, depending on the working capacity of the plant.

Storage accommodation should always ensure proper air circulation, and strictly avoid any stagnation or humidity, or very sudden variation in air temperature. Cones or fruits should never be placed directly on concrete floors: wooden or brick floors are recommended. Windows should be without glass but protected by shutters. Nonmetallic roof covering is preferable.

A many-storied building is usually better as elevators or other equipment allow easy transportation between different floors. Floor surface should be divided into single compartments each containing fruits or cones of a single provenance; to assure an easy separation of provenances, the area of each compartment should not he more than 8 to 10 square meters. Cones or fruits should be stored in thin layers (no more than three or four times their length) and stirred very often. Sometimes a part of the compartment floor is easily moved (or opened), so that fruits may fall down to the floor below.

The total area of the store depends on the amount of fruits to be handled, on the duration of storage, and on the working capacity of the seed extractory. If many species are worked in the same plant, one should bear in mind that their collection times are often different: the storage surface could then be reduced. In central Europe, for example, spruce cones are collected in September-October, while those of larch and pines are collected much later, sometimes even in winter. By the beginning of the collection time for larch, a large proportion of the spruce cones will have been already handled, and the space in the store emptied.

Techniques for Extraction

The separation of the seed may be obtained by purely mechanical means, such as threshing; by maceration (fleshy fruits); or by drying the fruit. The last is the more frequent case, almost general for conifers, whose cones should be dried to get their scales open and release the wed.

Drying

All fruits (including cones), have at their full ripeness a variable but always high moisture content. For example, in cones of some pines this varies between 35 and 65 percent. In natural conditions, fruits on the trees are largely exposed to heating by the sun, and to open air movement; both, but especially the second agency, provide a progressive drying of the fruit, which eventually causes the splitting of the fruit envelopes (i.e., beech or chestnut), or the opening of the cone scales (i.e., *Pinus* spp., *Picea* spp.), or the disintegration of the whole cone (i.e., *Abies* spp., *Cedrus* spp.). The speed of this process depends upon the climatic conditions and the local weather, and its duration varies greatly according to the species.

Cones of Abies disintegrate just as seeds are ripe, while the same process occurs in *Cedrus* spp. only some months later. Spruces open their cones and seed falls early in autumn, while cones of larch and of many pines remain closed till late winter. Of course any variation of the local weather can change the atmospheric conditions and induce cones to open. In foggy, rainy autumns spruce cones too remain closed till very late, but a few dry, windy days make them open quickly.

Artificial drying should reproduce as much as possible the natural process. Fruits should be submitted to progressive drying by a continuous release of moisture from the fruits. Air coming in contact with fruits should be always drier than the fruits. This can be obtained by circulation of air, either artificially heated or unheated.

The circulation of air at room temperature is sufficient to open and disintegrate cones of *Abies* and *Cedrus* spp., and separate seeds of some hardwoods, such as beech or oak, from their hulls or coverings. Fruits, which are placed in thin layers on the floor (which should be of wood or bricks, not of concrete) or on canvas, in a well-ventilated room, should be stirred frequently to avoid any moisture stagnation.

The stores described for keeping cones or fruits could very well be used for air-drying too. This method is used for very small amounts of fruits, which do not warrant any special equipment, and for species which could be damaged by any heating, including sun heating. It is commonly used for *Abies* spp. Depending upon the atmospheric conditions, its duration is often very long, and in a wet season a good deal of work is called for in stirring cones frequently.

Drying in the sun is still largely used. It is convenient for occasional operations, but for some species it is commonly used for regular operations too. This was of course the first method followed in fruit drying. Fruits are placed in layers on platforms or on screens, and exposed to the sun. Mediterranean pines, such as P. pinea, and P. halepensis, are always handled by this method. Medium-sized cones, such as those of P. pinaster or P. halepensis, are placed on a platform (of pressed soil, or bricks) in a uniform layer 10 to 15 centimeters deep, stirred frequently, and exposed to the sun until scales open; a cover is spread over the cones in bad weather.

As cones open, they are shaken on the platform, and seed is collected on it. Cones of P. pinea are placed in long rows, piled about 50 centimeters high. Open cones are gradually collected from the upper layer of the row and are ready for tumbling. Other species are exposed to the sun on wire screens with meshes of suitable size to let seeds drop through onto a canvas spread under the screen.

If local climate is quite cold and damp at night, screens should be removed and put under cover. A glass is sometimes placed over the screen, to protect cones against rain and to avoid the dispersal of sun heat; but this method is rarely used today since it requires too much labour, and glass may cause overheating of the cones with damage to the seed. Cones of most species generally open in the sun in a relatively short time: white and yellow pine and Douglas fir require from three to eight days; but other species could require more than one month. Some species such as longepole pine, open so irregularly in the sun that artificial drying is recommended.

Drying is easy during spring and summer, but during autumn and winter when most cones and fruits ripen, it can be carried out only in hot dry regions. There, it is often very convenient, since it does not require expensive installations; but where the weather is changeable, artificial heating is more convenient, as it permits normal work irrespective of weather conditions.

Drying in heated kilns is used mostly for cones of conifers, and therefore only cones will be dealt with here, although some other fruits can be handled in this way. The main difficulty in drying cones by natural methods is the absolute impossibility of controlling air humidity and temperature. An increase in the humidity of the air may cause a reclosing of the cones and sometimes a consequent "case-hardening," i.e., cones do not react to further drying but remain closed.

Artificial heating, on the contrary, permits control of air moisture and temperature, a much shorter handling time is required, and the process can be continuous so that work organisation becomes more effective. Of course, artificial heating requires installations and equipment which are often very expensive. Permanent extraction plants with such equipment should therefore be established only if enough seed is obtained to make the investment pay.

Small drying units mounted on trucks have been tried in the past. The first one was constructed at, Eberswalde in about 1930; this unit could be moved from one forest to another, and could extract separately the seed of each local provenance, thus bringing about some reduction in the transportation cost. Recently, infrared equipment has been tested to dry cones on the truck. But permanent units are on the whole more effective and seem to be more economical.

Cone drying by artificial heat should be carried out in such a way as to obtain drying in the shortest time, without damaging the viability of the seed. For this purpose, the following recommendations should be carefully observed.

1. Cones should be properly cured before they are put in the kiln.
2. Air temperature should be controlled and kept to the minimum level sufficient to dry cones.
3. Cones should not be heated longer than necessary.
4. Air in the kiln should be kept as dry as possible.

Moisture and temperature controls are of the utmost importance. The ideal handling would be to expose the cones to a flow of air, whose moisture content should decrease gradually, thus getting a regular, progressive drying and opening of the cone scales.

Heating increases the saturation deficit of the air, thus obtaining a further release of moisture from cones. But air temperature must not rise above a certain limit, beyond which seed viability in seriously damaged. Generally, temperatures of 40°C to 50°C may be reached. Such temperatures, however, are only allowed if the air humidity and the cone moisture are very low, a combination occurring only at the end of the drying process. At the beginning, hot air acting on wet cones would cause scalding of the seed.

In the past, decreasing temperatures have occasionally been used: but in modern extraction plants either constant or increasing temperatures are employed.

Moisture control could be obtained by chemical means, forcing air through filters containing drying materials, but heating is generally much easier and cheaper and it is almost always preferred.

Air heating: This can be obtained by different methods, any source of heat being suitable. In large extraction plants any kind of fuel can be used, including emptied cones or other wastes of extraction. Modern plants often use oil, since it insures a more regular combustion and a completely automatic temperature control. Sometimes electric heating is employed, but it is very expensive and is therefore used only for very small dryers, e.g., for laboratory units.

The control of air temperature is relatively simple in modern plants. A thermostat located where the air enters the drying room can stop or slow down the heating, or permit a flow of cold air to enter and reduce temperature. In old plants temperature is controlled by the operator. Thermometers and thermographs allow the operator to follow the variation of temperature inside the kiln; they should be connected to an alarm bell which rings when a critical temperature is reached.

Air humidity: The control of air humidity is much more difficult: hygrometers and hygrographs can record the humidity variation, but it can be regulated inside the kiln only by forced ventilation. As already mentioned, seed may also be damaged by relatively low temperatures when humidity is high, so that every possible care should be taken to avoid any stagnation of moist air in the kiln.

Modern extraction machinery is generally based on a continuous forced ventilation, but in the case of old types of extracting machinery the air change is only periodic or sometimes completely absent. According to Toumey and Korstian, the relative humidity in the drying room should be kept lower than 50 percent during the early period of drying and below 10 percent at the end of the handling.

Precuring

Cones or fruits are collected as they appear to be ripe, but since natural seed dispersal often immediately follows full ripeness, collection starts before ripeness, particularly if a large amount is required. Fruits do not all ripen at the same time, however, even for the same species and in the same forest. Consequently, many fruits or cones are not fully ripe when they reach the extractory. They can be ripened by storing in thin layers and in well-ventilated rooms.

The longer this precuring period lasts, the more complete is the cone ripeness and the better is the seed quality. It has been found that the shortest time to get good results is two weeks, but many species demand more than six or eight weeks. Precuring avoids the case-hardening of cones which lose from 10 to 50 percent moisture according to the length of the precuring period, thus facilitating and shortening the drying process in the kilns. This saves fuel and considerably reduces handling costs. Just before the cones are passed into the kiln precuring can be carried out effectively by setting the cones in a local store situated near the top of the kiln, where air from the exhaust, still warm, passes through.

In the past many types of kiln of local design and construction have been used but they were often only suitable for local conditions. Modern installations present a large variety of types, which fall broadly into two major categories: progressive kilns and rotating drum kilns.

Progressive kilns: In this type of kiln, cones are exposed to an air flow getting hotter and hotter, and drier and drier, from the beginning to the end of the drying period.

Schematically, the cones are placed in the kiln on a set of trays, one above the other, in a sort of tower; trays descend during cone treatment. From the bottom of the kiln the hot dry air hits the lowest tray, containing fairly dry cones; passing through the cones, the air loses some heat and takes some moisture from the cones. In the second tray, from the bottom, cones have somewhat higher moisture content and the air is less hot and less dry.

In the last tray, the cones, still having the original moisture content, are engulfed by warm, humid air. At the right moment, a tray is taken away at the bottom, and the whole tray set descends one step, while a new tray with fresh cones is put on the top of the pile. Hot air can circulate simply by convection, but forced ventilation makes the operation quicker and more regular. The distance between two successive trays should be sufficient to contain open cones which have a volume 2 to 3 times that of closed cones.

The time required for complete drying varies according to the initial moisture content of the cones, the volume of circulating air, and the type of air circulation (convection or forced ventilation). For some species (i.e., Abies spp.) heating is harmful, and drying is obtained by forced circulation of air at room temperature. In some kilns, cones are forced to fall down from one tray to the one below by opening the trays bottom, or by inclining the tray. In other types, trays are replaced by slow moving bands: cones move on a band till they fall to the one below.

In the last two types, cones are shaken as trays fall from tray to tray, so that most of the seeds come out: special devices under the bottom tray collect the seed, while cones pass to further handling. Similar working schemes are applied in some small extraction units, planned chiefly for laboratory extraction. The Forest Experiment Station of Mariabrunn (Austria) has developed a small kiln of this type to extract seed from cones of individual trees, as well as to carry out experimental seed extraction to test the seed yield. This type usually has electric power heating.

Some kilns have just one or two trays, which may be opened: from the lowest, cones fall into a drum, where they are tumbled while still drying. This is the method used in some old and ver N, large extraction plants, still operating satisfactorily. At Wolfgang (Germany) four such "towers- could extract all the conifer seed required in the Federal Republic.

Rotating drum kilns: Cones are put in a cylinder made of perforated steel plate, rotating on a central axis. The cylinder is contained in a box, where forced air circulation is provided by an electric fan. During the extraction the cylinder rotates continuously, shaking the cones. The air temperature increases gradually, from room temperature to the maximum fixed level. The strong ventilation, the progressive heating and the continuous shaking make the cones dry and open in a short time.

Seeds released from cones pass through the holes of the plate, and are blown at once out of the kiln. Temperature and ventilation control is generally fully automatic. This type of kiln offers many advantages, and is usually preferred in modem extraction plants. The time needed for seed extraction is greatly reduced; the shaking of cones during drying causes the cones to open and the seed to fall. Separate tumbling is unnecessary, and labour and time are saved.

Seed is taken out of the hot air as soon as it comes out of the cone in order to avoid seed damage. The capacity of the cylinder is reduced so that extraction can be done separately for cones from different provenances, even if the quantity is small. Kilns of this type may have electrical heating or a separate heating unit which can use different fuels (cones, coal, oil).

The cylinder, the electric motor, the ventilator and the equipment for air and temperature control are contained in a steel box above which is a predrying box with two floors. From this box, cones fall into the cylinder, where they are exposed to increasing temperatures (40-45-50-55-60°C) for three to four hours. At the end of the extraction, reversing makes the cylinder open: emptied cones fall down, and other cones come into the kiln. This type has a working capacity of 25 to 30 hectoliters of cones per day.

Modern rotating kilns are generally of compact metal construction, moderate in size so that they can be placed in small, inexpensive buildings. The output is relatively low as the capacity of the cylinder is of 3 to 4 hectoliters. These kilns are generally planned for working in a set of two or three, so that they are suitable for plants of different sizes. Their moderate cost is also a factor which makes them suitable for use in small plants.

Occasional drying methods

For small amounts of cones, and especially for occasional operations, drying can be obtained by very simple means. A room with any source of heating is sufficient. If possible, cones should be spread on single trays, and placed near, but not too close to, the source of heat. Doors and windows may be opened carefully or appropriate holes made in them to ensure proper air circulation. Heating should be kept low and cones should be frequently stirred.

When such primitive treatment is adopted, any haste may cause damage to the seed, because safe drying can be obtained only by a very limited heating and therefore takes quite a long time. Kilns constructed for other purposes, such as lumberdrying or hop-curing kilns, can be used for cone drying cheaply and with good results. Of course, care should first be taken to see what facilities they have for the control and regulation of temperature and humidity.

Any artificial heating involves, of course, a fire hazard. This is particularly true in handling cones, which always contain resin, often in large amounts; besides which, the cone scales themselves are, when dry, very inflammable. Great care must therefore be taken. No heating of any kind should be allowed in the cone store. In the extractory plant dust, which consists chiefly of particles of cones and resin, should be removed, especially near the kilns. In modern plants, this can be done by a vacuum equipment.

Pneumatic pipes suck the dust from the interior of the machines into a special container. The building where drying equipment is installed should be of masonry, and wood should never be used in the construction of the heating and kiln compartments. All measures should be taken to prevent and control possible fires, such as the provision of extinguishers and the instalment of iron doors.

Tumbling

When cones open after natural or artificial drying, not all seeds are detached from the scales and fall out; some are usually left inside the cone. Seeds of some species fall easily after a thorough shaking, but cones of others require a severe and prolonged jarring. The method by which cones have been dried makes a great difference. Adequate precuring, regular drying, and no excessive heating ensure a more complete opening of the cone scale and an easy seed release.

After drying, especially by heating, the seed should be removed from the cone as soon as possible, because open cones exposed to cold, wet air, can reclose in a short time. It would be wise to keep dried cones in a warm room until all seed extraction processes have been carried out. In order to separate seed, cones are put into a drum or tumbler, which is a barrel or cylindrical or rectangular box, rotating on a central axis; the walls of the barrel or box are generally of wire mesh large enough to let seed pass through. As the drum revolves, the cones are thrown against the walls, more or less violently according to the revolving speed, while the seed comes out of the cone and falls through the wall meshes into a collector.

The process is carried out in the same way as in the cylinder of rotating kilns. The drum may be closed at both ends, and filled and emptied at the end of each operation cycle. In modem types, reversing makes the cylinder open; in other types the drum is open at both ends and is quite long, with its axis somewhat inclined so that the cones are fed in at the upper end and, during rotation, roll slowly to the other end, where they fall out. Such a drum gives a continuous operation cycle, but the rotation speed must be slow, with the result that the vibration may be inadequate for some species.

The walls of the drum are sometimes of thin bars, placed very close together. All such drums can be operated by hand or by a motor; the turning speed can vary according to the species. Small types are easily transportable, and can be used for very small amounts of cones in occasional operations. In these cases, some devices similar to those used for agricultural crops may also give satisfactory results.

If no equipment is available, seed extraction can be obtained by flailing open cones on a floor; in the case of very small cones, they are put into sacks before flailing. For species with very small Seeds, such as *Tsuga* and *Cupressus,* a little revolving box is used: light wooden frames support a mosquito wire net. The box rotates on its axis, just as in larger tumblers, but for these species a few rotations suffice to extract all the seed.

Threshing

Seeds of some species, such as *Liriodendron, Catalpa* and *Cercis,* can be extracted easily from the fruit. Their pods are flailed on a platform and generally break without difficulty. For large quantities mechanical threshers are used. Generally, threshers used in agriculture for similar types of fruit can be easily adjusted by altering the distance between the crushers. The other operations follow in the usual way. The cones of some conifers, especially some pines, do not open completely even after a long drying, and even tumbling is not sufficient, so that further mechanical handling is needed to extract the seed still remaining in the cone.

After sun drying, cones of *P. pinea* are treated in large threshers, so that the scales detach quickly and completely from the cone axis. Larch cones are often difficult to open as they have a very high resin content and the resin melts at a temperature lower than that sometimes required to open the cone. After partial drying, therefore, the cones are crushed in special mills.

In stationary or rotating kilns, the cones, after normal drying, fall into a cylinder equipped with a series of knives: fast rotation provides both tumbling and cutting of the cones at the same time. For some fruits, hammermills are used, always with the aim of breaking the casing of the fruit. Again, hammermills used in agriculture are quite effective and suitable.

The processes of tumbling; threshing, crushing or cracking can cause damage to the seed, thus inducing an abnormal seedling development or an early loss of the germination power. These mechanical processes should be carried out carefully, the intensity of shaking and flailing being proportionate to the hardness of the seed walls. It is better to leave some wed inside the cones than to risk severe damage to the seed.

De-pulping

The fruits of many species, such as *Juniperus* spp., *Prunus* spp., *Malus* spp. and *Rhamnus* spp., are fleshy or semifleshy. Some of these fruits (i.e., *Juniperm* spp.), may be dried just before storage, drying being effected by natural or artificial heat. But safe storage usually requires separation of seed from the pulp of fruit: this can be obtained by macerating the fruit in water. Fruits are left in open containers until the pulp becomes soft and decays; they are then placed in water (often some lye is added), stirred and mashed.

After a certain time, the seed separates from the pulpy parts, and sinks to the bottom: after proper drying (preferably by unheated air) the seed finally goes to cleaning or to storage. Some special machines have been developed for de-pulping fleshy fruits and adapted hammermills, cornshellers and macerators have also been used. A small portable type of macerator consists of a plate spinning at the bottom of a seed hopper: pulps are washed out through an appropriate outlet, while seed is retained.

Cleaning

Many fruits are quite ready for sowing as they are collected, or just require the separation of seed from the casing. Beech, oaks, chestnut, and hickory require only a very rough cleaning, which can be done by winnowing, screening, or some other very simple method. These seeds are often extracted by hand, or handled in a separator normally used for agricultural products. Almost all the abovementioned machinery can do an effective job provided the size of the seed is taken into account.

Seed which has been extracted by the various methods already described requires further handling before storage. Wings or some other parts of he fruit may be still. attached to the seed, and many impurities of different types are always present. Cleaning provides a thorough separation of pure seed from inert matter and from shriveled and empty seeds, and prepares it for storage and sowing in the beat condition. The species and the amount of the seed will determine whether cleaning is best carried out by hand, by improvised equipment, or with the aid of special machines.

De-winging, fanning, screening, and separation are often needed to obtain seed of good quality. As a general rule, the aim of cleaning is to get completely pure seed, but this is not always possible. On the other hand, there is often no particular advantage in cleaning beyond a certain limit. Modern machinery has reached a very high degree of efficiency in this field, but in many cases, complete separation would cause a loss of seed. In the case of larch, for instance, the seed should not be cleaned to a purity higher than 65 percent as it has been proved that further cleaning causes a drastic loss of good seed.

De-Winging

Many forest seeds, and almost all those of conifers, are

winged. In some species, e.g., *Pinus pinea*, the wing is quite rudimentary or very short, but in others it is long and hard. Sometimes the wing completely covers the seed (*Abies* spp.). For some species it would be preferable to store a seed with its wings in order to ensure a better air circulation. On the other hand, sowing winged seeds may cause many difficulties: they may be blown away or more easily picked up by birds, and the use of sowing machines may be impossible. For these reasons, the wings are always removed whenever they are larger than the seed.

De-winging is almost always carried out by rubbing. In the past, seeds were spread on a floor and sometimes lightly moistened by sprinkling them with water, after which, flailing by light leather flails was sufficient to remove wings. This method is no longer used, because it involves the risk of fermentation of moist geed or the development of fungi. Hand rubbing the seed in a sack is the safest method, but it is practicable only when small amounts of seed are involved.

Many types of de-winging machines have been developed and are in use, from the simplest types, which are operated by hand, to those suitable for large plants, which work semi-automatically, and give a continuous output. They are almost all rotating devices. Rotatory de-wingers may have brushes or knobs. In the first type seed is pressed by brushes moving against the walls of a cylinder. In the second type rotating knobs or pads force the seed to pass through narrow outlets, and leave the wings.

The distance between brushes or knobs can be regulated according to the size of the seed. Brushes may be of steel, nylon or tampico, which have different degrees of hardness; cylinder walls may be of corrugated wire or perforated plate, or lined with canvas or rubber. A de-winging machine may be a drum of webbing revolving in a hammock of the same material. Another de-winger is made with rubber pads rotating in a rubber cylinder.

If the clearance between knobs or brushes and the walls is too narrow, or if the brushes are too hard, the seed may be damaged either by pressure which breaks the seed coat; by scarification; or by heating caused by friction. Great care should be taken to avoid this. The main difficulties to be overcome are: a proper regulation of the clearance; the settlement of the best revolving speed; and the attainment of a regular seed supply. For some very sensitive seeds, cleaning before dewinging is recommended, to remove any particle of hard material which could injure the seed coats during handling.

Cleaning operations

The different cleaning operations may be carried out separately or at the same time. Modem cleaning machines often combine different handling methods so that the cleaning process may be both effective and quick. These operations are given below.

Fanning or winnowing: A strong air current blows away fight impurities, cone particles, dust, empty seeds, while good seed and heavier particles are not removed or fall down at once. Winnowing is often carried out, especially in agriculture, by throwing seed into the air on windy days. In winnowing machines the air current may be either horizontal and blow through the seed which falls gradually, or vertical and blow across a moving screen on which seed passes.

Screening: Largely used in many separating operations: by passing through screens with meshes of different sizes, the seed is separated from impurities which are larger or smaller than itself. Screens may be of metal net or of perforated plate. Shaking screens are often used. Screens generally work in series of different mesh or hole size.

Separation by centrifugal force, or by vibration: If seed is centrifuged in an open cylinder, heavier particles go up first along the walls and escape: variation in rotation speed provides progressive separation of the particles of different weight. The same effect is obtained by vibration.

Flotation: When put into water, good seed sinks, while light impurities and empty seed float on the surface and can easily be skimmed off. This simple separation method requires a subsequent seed drying, but it is very effective with some species, such as acorns.

Combine cleaners: Cleaning methods separate good seed from undesirable impurities or empty seeds in three ways: by size (screening); by specific gravity (flotation, vibration); and by centrifugal force. These different methods are often combined in modem equipment. Screening and winnowing are almost always used, and sometimes carried out simultaneously. There are many such machines. Mechanical shaking and separation by air blowing are used, a vibrator type seems to be especially effective in cleaning larch seed. For hardwoods, fanning mills have been developed. Vacuum and cyclone systems have been successfully used in cleaning seed, but up to now such special handling methods are rarely used.

Conveyors, baggers, and other transport equipment inside the seed extractory

All material, from the rough fruit or cone to the fully cleaned seed, should pass from one handling to the next quickly, in order to avoid any possibility of contamination or mixture with undesirable matter, and any remoistening which would necessitate further processing. Transport should be done as far as possible by mechanical means: conveyors, transport bands and blowing pipes should be used in the different phases of handling, depending on the size and the type of material. In general, cones are transported by chain elevators or by bands; for seed, blowing by pipe is very convenient. These blowers should be self-cleaning, in order to avoid any possibility of mixtures, chiefly between seeds of the same species but from different provenances.

All handling, but especially tumbling and cleaning, produces a large amount of dust. De-dusting apparatus should be connected to all transport equipment, as well as to all handling machines. Vacuum pipes should collect all dust from where it is produced, inside the kiln, inside other machines or at their air outlet, and convey it to a cyclone and a special cell. This keeps the work room clean all the time and makes working conditions safer. Transport equipment should carry cleaned seed up to the store. Packaging into containers for storage or for shipping could also be done by automatic bagger.

Moisture control in drying

Seed should be kept dry during all the different stages of handling, but particularly at the end of the process, where it is prepared for storage or shipping. Seed handling should be carried out in a warm, well ventilated room. In effect, many cleaning machines use the heated air coming from the kilns, thus Insuring a continuous drying process. But, in any case, before storing or shipping seed, moisture should be checked and especially if the seed is to be stored in closed containers.

Testing the actual seed moisture is rather difficult. Laboratory moisture testing may give very exact results but always takes rather a long time. Furthermore, the specialised staff required is not normally available within the seed extractory. In practice, testing should quickly give the moisture content of seed at the moment of packing. For this purpose, some electronic moisture testers have been developed, based on the electric capacity or resistance; some types give the degree of moisture of seed inside the container.

Infrared moisture balances are also used. All the above mentioned testers give results by direct reading but their precision is somewhat limited: further development of such apparatus is desirable. In storage, seed should have a low moisture content (generally of 5 to 10 percent); if necessary, it should be carefully dried by submitting it to a dry draught of air. For this purpose almost all types of stationary kilns, or even a simple winnowing machine, should give immediate results. Temperature should be kept very low, however, and for some species (i.e., *Abies* spp.) any heating should be avoided.

9

Seed Treatment for Agronomic Crops

Most seed treatment products are fungicides or insecticides applied to seed before planting. Fungicides are used to control diseases of seeds and seedlings; insecticides are used to control insect pests. Some seed treatment products are sold as combinations of fungicide and insecticide. Fungicidal seed treatments are used for three reasons:

(1) to control soil-borne fungal disease organisms (pathogens) that cause seed rots, damping-off, seedling blights and root rot;

(2) to control fungal pathogens that are surface-borne on the seed, such as those that cause covered smuts of barley and oats, bunt of wheat, black point of cereal grains, and seed-borne safflower rust; and

(3) to control internally seed-borne fungal pathogens such as the loose smut fungi of cereals (Figure 1).

Most fungicidal seed treatments do not control bacterial pathogens and most will not control all types of fungal diseases, so it is important to carefully choose the treatment that provides the best control of the disease organisms present on the seed or potentially present in the soil. The degree of control will vary with product, rate, environmental conditions and disease organisms present. Some systemic fungicidal seed treatments may also provide protection against early-season infection by leaf diseases.

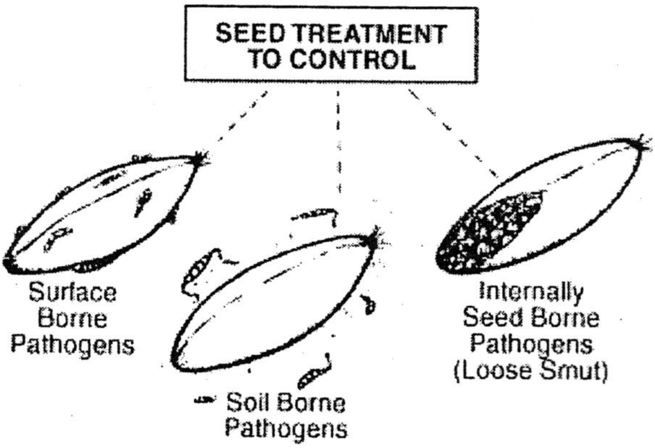

Figure 1. Reasons for seed treatment.

Fungicide-insecticide combination products or an addition of insecticide for wireworm control should be considered if planting newly opened land or land that has had a history of wireworms. Consult current recommendations for insecticides registered for wireworm control. A fungicide-insecticide combination also may be useful for dry beans. The insecticide used on dry beans should be one that provides control of the seed corn maggot.

Fungicidal seed treatments help control soil-borne pathogens that cause seed decay, seedling blight and root rot. Control of these diseases may result in better stands, more vigorous seedlings, and increased yields. Protectant fungicides such as captan, maneb, PCNB, thiram, or fludioxonil (Maxim) help control most types of soil-borne pathogens, except for common root rot and take-all. Protectant fungicides containing captan, maneb, or thiram are sold under various trade names.

Seed Quality

A number of factors affect the quality of seed. Seed-borne fungi, insect feeding damage, poor seed storage (temperature and humidity too high), and damaged seed coats can reduce the quality of seed. Seed quality may be reduced due to infection by fungi when harvest of the seed crop is delayed several weeks or when seed diseases occur due to wet conditions. Infected seed can be detected at harvest as discoloured, shriveled, or moldy in appearance. Insect feeding that may occur in the field prior to harvest or in storage may affect seed quality. Germination of seed lots may be reduced by the insect partially eating the seed or opening the seed coat to fungal infections occurring at a later date.

The seed coat is the primary protective layer surrounding the seed. Seed coat damage can result from rough handling during harvest, and cleaning or planting operations. Small cracks in the seed coat increase the chances of seed-rot by permitting water-soluble nutrients to escape into the soil which activate soilborne fungi that quickly enter the damaged seed. Grain intended for seed should be handled as little as possible, dried using low temperature drying, and stored at moisture levels below 13 percent. High humidity and temperature during storage promotes growth of fungi and insect damage. Seed lots should be closely monitored for mold growth and insect activity during storage.

Importance of Seed Treatment Pay

Seed treatments can play an important role in achieving

uniform seedling emergence under certain conditions. The selective use of seed treatments can protect seeds or seedlings from early season disease and insect pests affecting crop emergence and growth. A few seed treatments may also be used to enhance crop performance during the growing season. For example, treatments containing Rhizobium inoculant have commonly been used to enhance the nitrogen fixing capability of legume crops. Increased germination can be observed when seed treatments containing fungicides are used when poor germination results from a fungal infection.

Fungicide are available to protect seed and seedlings from many seed-borne or soil-borne pathogens. Seed treatments are most beneficial for seeds infected prior to planting or when cool, wet soil conditions exist at planting which results in delayed germination. Seed treatments containing insecticides can serve to protect seed and seedlings from certain insect pests found in the soil. Cool, wet, seed beds resulting in delayed germination or plowing down green manure crops may warrant including an insecticide seed treatment.

Seed treatments should not be considered a cure-all for the selection of poor seed lots. Seed treatments will not increase poor germination due to excessive mechanical damage, poor storage conditions, genetic differences in variety, or other damage. The effect of a seed treatment on plant stands will depend upon seed quality as well as field conditions at planting. Extensive research has shown seed treatments can increase stands from poor quality seed or when planting conditions are less than optimum. However, seed treatments do not always increase yield. Generally, seed treatments result in increased yield by preventing death of seedlings causing frequent skips within crop rows.

Grain producers should realise that any one seed treatment will not control all diseases or insects that may attack seeds and seedlings. Protection is also short lived, generally only lasting as long as it takes for the crop to emerge. Before selecting a seed treatment it is essential for the grower to consider potential problems associated with the seed and the history of problems associated with each field. Each planting situation is different requiring consideration of seed quality; planting rate; tillage and seed bed preparation; soil moisture and temperature; time of planting; and the likelihood of rapid emergence.

The purpose of laboratory testing of seed germination is to assess seed quality or viability and to predict performance of the seed and seedling in the field. Seed processed for sale must be tested by a qualified laboratory under the Association of Official Seed Analysts Rules for testing seeds. Several different kinds of testing are available depending on the type of seed to be tested, the conditions of the test, and the potential uses of the seed. The most common tests are the warm germination test, cold germination test, accelerated aging test, and the tetrazolium test.

Each test is designed to evaluate various qualities of the seed. Factors that can affect the performance of seed in germination tests include; diseased seed, old seed, mechanically damaged seed, seed stored under high moisture, and excessive heating of seed during storage or drying. In most cases a seed treatment will improve germination of seed only if the poor quality is due to seed-borne disease. The most common test is a warm germination test because it is required by seed laws to appear on the label.

The percentage of germinating seed in a warm germination test must be printed on the label of the seed if it is to be sold as seed. Germination is defined as: "the emergence and development from the seed embryo of those essential structures which are indicative of the ability to produce a normal plant under favourable conditions." The warm germination test reflects the stand producing potential of a seed lot under ideal planting conditions. Usually 400 seed from each seed lot are placed under moist conditions on blotters, rolled towels, or sand and maintained at 77 degrees F for about seven days.

At the end of this period the seedlings are categorised as normal, abnormal, or diseased, and dead or hard seeds. The percentage germination is calculated from the number of normal seedlings from the total number of seeds evaluated. The cold germination test is designed to measure the ability of seeds to germinate under high soil-moisture content and low soil temperature. This vigour test simulates early season adverse field conditions and usually represents the lowest germination that would be expected from a seed lot planted under such conditions.

Actual field germination would normally fall between the cold test result and that of the warm germination test. Seeds are planted in a sand-soil mix at high moisture content and maintained at 50 degrees F for seven days. The test is then placed at 77 degrees F for four days. The percentage of healthy seedlings that emerge at least one inch above the soil is reported. The accelerated aging test estimates the carryover potential of a seed lot in warehouse storage. The seeds are exposed to high temperatures and high relative humidity for short periods of time that cause seed deterioration.

Seed are suspended over water in a chamber for 72 hours (wheat and soybeans) or 96 hours (corn) then tested in a standard warm germination test. This test only would be used on seed whose longevity was in question. The tetrazolium test is a "quick test" for seed lot viability. It is useful when an approximate germination percentage is needed immediately. Seeds are soaked overnight in water then treated with tetrazolium to give an indication of viable, abnormal, and dead seeds in the seed lot. This test will not detect seed-borne disease, thus is limited in its ability to estimate seed quality. This test is highly reliable for determining viable seed of corn, wheat, oats, barley, and other grasses.

If growers wish to use bin-run seed and has not had a germination test conducted by a competent laboratory, they can get an approximate germination test using the following procedure in their own home. Place two paper towels in the bottom of a dish or tray, one on top of the other; wet the towels thoroughly and tilt the tray up on one end so that excess water runs off the tray. Select a random sample of 100 seeds from the seed lot and place them in between the moist paper towels. Put the tray in a plastic bag and tie the end shut to prevent the towels from drying out.

Place the tray in a location of diffuse, not direct light, such as a north window. The location should be warm enough (75 to 85 degrees F) for good plant growth. An ideal location is with well growing house plants. After five to seven days, open the plastic bag and count the number of germinated seed with intact tap roots and shoots. Do not count moldy seed or diseased seedlings. Testing 400 seeds in this way will give a good indication of the germination percentage.

Fungicide

The basic component of most seed-treatment products is a fungicide. Insecticides are contained in some products to give insect protection when such control is desirable. The grower should select a seed treatment product that will control the most likely pest to be encountered within a given field. It is important to note that not all fungi are alike. One fungicide seed treatment may be very effective against one fungus but will have no effect on another.

Rates of active ingredient used also affect a seed treatment performance on different disease problems. Fungicides have fungicidal activity, that actually kills the fungus, or fungistatic activity, that slows the growth of the fungus to prevent extensive damage. Fungicides used in seed treatment are of three types: protectants that generally prevent infection of seeds and seedlings by soilborne fungi, surface disinfestants that kill fungi and fungal spores on the surface of the seed coat (seed-borne fungi) and "systemic" disinfectants that kill fungi already established within the seed (seed-borne fungi).

Insecticides used in seed treatments act as surface protectants killing insects as they feed on the seed or emerging seedling. The marketing of seed treatments by manufacturers may result in the same chemical being sold under several brand names and by several different companies. All chemicals have three names on the packaging; chemical name, common name, and trade or brand name.

The chemical name represents the molecular structure of the active ingredient and is often long and complex. The common name is used by the manufacturer in lieu of using the longer chemical name. The trade or brand name is used by the manufacturer in product marketing. For example, the chemical name (R,S)-2-[(2,6-dimethlyphenly)-methoxyacetylamino]-proprionic acid methyl ester has the common name mefenoxam and is marketed under the trade or brand name Apron XL LS.

Using mixtures of different seed treatment fungicides or insecticides is a common practice. Mixing two or more fungicides or insecticides results in a product that will control a broader spectrum of seed pests.

Application of Seed Treatments

Seed treatments come in a variety of formulations: dry flowables (DF), liquid flowables (LF), true liquids (TL), emulsifiable concentrates (EC), dusts (D), wettable powders (WP) and others. These materials may be used in slurry and mist type seed treaters upon mixing with water as specified in instructions on the product label. Special adhesives are incorporated into the dust formulations to adhere pesticide particles to the seed surface. Always apply seed treatment materials according to label instructions.

Commercial Seed Treatment

Commercial seed treatment is often desirable due to the specialised equipment required to properly apply treatments or to treat large volumes of seed. The primary concern of the commercial treater is equipment calibration to ensure delivering the proper amount of active ingredient on the seed. This has become especially important with more modern fungicides that require only very small amounts of material (less than 0.5 oz) per hundred weight of seed.

Detailed information on the process of treater equipment calibration can be obtained from the equipment manufacturer. Seed treatment manufacturer labels provide information related to mixing formulations and mixing procedures. Information needed to properly calibrate a seed treater includes:

1) the correct labelled rate of the seed treatment material that will be used,

2) how many pounds of seed is dumped each time the weigh pan arm dumps and,

3) the size of the cup being used in the treater.

This information will allow you to calculate the correct application rate for the seed and treatment material used. A new generation of seed treater is becoming available that offers precision distribution of treatment chemicals on individual seed through atomisation. These advanced controlled droplet application (CDA) machines offer complete seed coverage with lower water volume, shorter drying time, and continuous seed flow. These machines also reduce chemical exposure to employees and are capable of applying current and future products that may include both chemicals and biologicals.

On-farm Application

Many seed treatment materials are available for on-farm use. These are known as hopper-box or planter-box treatments. Most hopper-box treatments are dry treatments formulated with talc or graphite which adheres the treatment chemical to the seed. Liquid hopper-box treatments are available as a fast-drying formulation. Good seed coverage is required for maximum benefit from any seed treatment formulations. Obtaining thorough seed coverage can be difficult when attempting to treat seed in the field at planting time.

Most seed treatment companies recommend that the planter box be filled only half full with seed, and half the required amount of treatment product added. Mix the product into the seed volume with a stick. Then add the remainder of seed to the planter box and add the rest of the required amount of product, again mixing thoroughly. Good results are also obtained by premixing seed with the product in a suitable container, then pouring treated seed into the planter box.

On-the-farm treaters are available that apply liquid or dry formulations to seed as it passes through an auger from the transport bin or truck to the planter boxes. These treaters are a very convenient way to apply seed treatment onto bulk seed. For additional information on on-the-farm treaters, contact a representative of the company selling planter-box treatments.

Requirements for Labelling Treated Seed

Once seed is treated it can no longer be used for food feed or other purposes. Under the federal and state seed laws, the following information is required to be shown on the label: a word or statement in type no smaller than eight points indicating that the seeds have been treated; the commonly accepted, chemical (generic) name of the applied substance including the application rate, and a caution statement if the substance used in such a treatment has an amount remaining with the seed that is harmful to humans or other vertebrate animals.

Seed treated with highly-toxic substances shall be labelled in red ink to show a statement such as "Poison Treated." In addition, the label should show a representation of a skull and crossbones. Seed treated with less toxic substances shall be labelled to show a caution statement in no smaller than eight point type such as "Do not use for food, feed, or oil."

Colouring

All treated seed must be coloured an unnatural colour to distinguish it from untreated grain and prevent inadvertent use as food for man, feed for animals, or for oil purposes. All treated seed must be coloured with a dye, colour coat, or colour film which must cause no apparent injury to seed germination or danger to personnel processing or using the seed. Dyes are designed to stain the seed with colour. Colour coat pigments cover the seed with colour, and colour films are made from a polymer that creates a coloured film around the seed. Seed treatment companies currently offer a wide variety of colour and surface texture options to seed processors.

Safety Precautions

Seed treatment chemicals are pesticides and should be handled as such. Product labels provide information on safe handling and application. Be sure to follow all instructions as listed.

Hazards associated with handling, sorting, mixing, and storage of treated seed: Read the label and follow instructions carefully. Over treatment may injure the seed and under treatment will not control the pests in question. The label will also provide information on the toxicity of the active ingredients in the form of code words (Caution, Warning, Danger).

The label will provide basic information on first aid, environmental hazards, directions for use, proper storage and disposal of containers, and pesticide and general product information. Use the label and be familiar with it. Material safety data sheets (MSDS) are available from the company for each of the products sold. Consult the MSDS for more detailed information before applying the material.

When handling toxic substances, operating personnel must be provided with protective clothing such as coveralls, cap, protective glasses, rubber apron, rubber boots, rubber gloves, and a respirator designed for use with the substance. Personnel should not inhale the dust or vapour nor permit the material to contact the skin or eyes. The operator should wash thoroughly with soap and water before eating and smoking. Bathe immediately after work and change all clothing and wash clothing thoroughly with soap and hot water before reuse.

In case of contact, immediately remove contaminated clothing and wash skin thoroughly with soap and water. A safety shower should be installed in the immediate vicinity of the treater. An exhaust system should be installed to remove vapours and dust from the operating area. The exhaust air should discharge into a cyclone or bag-type dust collector. The treater should also be vented and tied into the exhaust system. Seed treaters should be isolated and not operated in the vicinity of other personnel or farm commodities that are to be used for food, feed, or oil. Special multiwall (3 or 4 ply) paper bags or tightly woven bags are recommended for seed that has been treated with toxic substances.

Seed should be thoroughly dry before going into the bag, as excessive moisture can cause rapid deterioration of the seed. Store treated seed in a cool, dry place away from food or feed products. Seed treaters should be thoroughly cleaned after use as some of the pesticides are corrosive and others settle out and cause clogging of the equipment. Do not run contaminated water into a stream or public sewer, but discharge into a shallow ground pit.

Hazards associated with the use of treated seed and the disposal of unused treated and empty pesticide and treated seed containers: The user of treated seed should read and follow the label carefully. The planter hoppers should be filled outdoors. Do not breath dust or fumes from treated seed or get in eyes or on skin. Wash thoroughly with soap and water after handling treated seed. Use treated seed as seed and do not introduce them into the food or feed channels because serious injury to livestock and humans may occur. Follow instructions on the label for disposal of seed treatment containers and bags containing treated seed. Do not reuse pesticide containers.

Triple rinse and, if possible, recycle containers. Containers can be punctured and disposed of in a sanitary landfill or incineration, or, if allowed by state and local authorities, by burning. If burning, stay out of the smoke. Old seed with low germination should be destroyed by burying at least 18 inches deep in an isolated area away from the water supplies. Treated seed exposed on soil surfaces will be hazardous to birds and other wildlife.

Biological Control of Seed Diseases

Recent research has focused on biological control of certain seed and seedling diseases of crops. The most promising biological control agents tested have been either fungi or bacteria. These biological seed treatments control seed pests by parasitising the pest organisms, competing for food on the root system, or producing toxic compounds that inhibit pathogen growth. Several biological seed treatments are currently on the market.

Insecticide Treatments

A limited number of insecticides are labelled for use as seed treatments to protect seedlings of various field crops from soil inhabiting insects.

Soil Insect Pests

Soil insect pests are generally more of a problem on corn and soybeans than on small grains and forages. Soil-applied insecticides recommended for control of rootworm larvae also may prevent early stand losses due to wireworms, grubs, seedcorn maggot, and seedcorn beetles. If a soil-applied insecticide is used at planting, then a seed applied insecticide may not be necessary. Refer to the label of the soil-applied insecticide to determine which insects are controlled before purchasing a seed treatment insecticide.

Stand losses to these soil insects is generally not significant enough to warrant routine use of soil-applied insecticides in first-year corn. The use of insecticide for first-year corn may change in the future due to a biotype of the western corn rootworm beetle. Benefits from suppression of early-season soil pests may be significant in 1) no-till or reduced-tillage conditions with poor weed control where a high risk exists for a given insect problem such as wireworms and grubs in corn following established sod, and 2) continuous corn where early-season pests may cause unacceptable stand losses.

Seedcorn Maggot

Seedcorn maggot or the bean seed maggot are small, yellowish-white, legless fly larvae found feeding on corn seeds. Extensive feeding by these maggots will cause a reduction in stand. In general, seedcorn maggot problems are most likely to occur in situations where;

1) high organic matter or decaying vegetation attracts egg-laying adult flies (i.e. a green cover crop such as rye or alfalfa is plowed or disked under), and
2) cool, damp soil conditions delay seed germination and prolong the period vulnerable to maggot attack.

No-till production systems have not been found to increase the likelihood of problems with seed maggots.

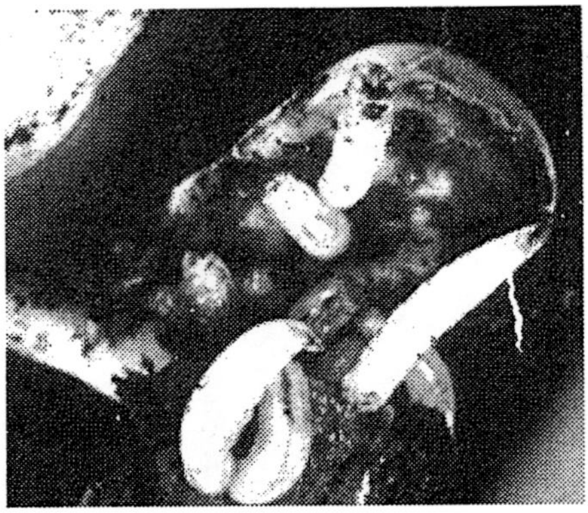

Figure 2. Seedcorn maggot

Several products are formulated for application prior to planting seed on the farm as planter-box or hopper-box treatments.

Seedcorn Beetle

Partially eaten seeds, loss of germination, or stunted seedlings in the presence of small (1/4 to 1/3 inch long) brown ground beetles indicate a seedcorn beetle problem.

Figure 3. Seedcorn beetle

As with seedcorn maggots, damage is most likely to occur under cool, damp conditions where seed germination and seedling development are delayed.

Wireworms

Wireworms are the larval stage of the "click beetles." The term "wireworm" applies to a complex of species with life cycles requiring one or more years per generation. Wireworm populations affecting corn are most severe in fields following sod or fields having a prolonged grassy weed problem. The larvae pass through a number of life stages, or instars. The earliest stages are very small and white, the latter stages have a characteristic hard-shell appearance and are yellow-brown in colour.

Full-grown larvae range from 1/2 to 1 inch long depending on the species. Wireworms damage corn by feeding on germinating seeds and young seedlings and may bore into stalks at the soil level. Stand loss may be significant in fields with high populations. In fields where a soil-applied insecticide is not used, especially first year corn fields, application of seed treatment insecticide is strongly recommended.

Treatment Time

Corn

Treatment is recommended for all seed corn to prevent or reduce seed decay and seedling blights. The germination and early growth of the seedling is a critical period in the life cycle of corn. Soilborne organisms may invade and kill the embryo before germination, or when invasion occurs later, the seedlings may be destroyed before or after emergence. Seedlings surviving fungal infections at germination are often less vigorous. Corn seedling diseases are more prevalent in cold, wet soil than when soil temperatures are above 55 degrees F.

Therefore, early planted corn needs the added protection of a good seed treatment fungicide. Seedling blights tend to be more severe in fields with poor seed bed preparation such as in no-tillage or reduced-tillage fields. Corn seed is generally treated with a fungicide or fungicide-insecticide combination by the seed producer or seed processor. Corn smut, leaf blights, stalk and ear rots, and virus diseases are not controlled by seed treatment fungicides.

Small Grains

Seed treatment of small grains is recommended to control smut diseases and to reduce seed decay and seedling blights. The smut diseases include stinking smut (common bunt) of wheat, loose smut of wheat and spelt, and the loose and covered smuts of oats and barley. Wheat and oat varieties have resistance to some, but not all, races of loose smut (wheat) and loose and covered smuts (oats).

The only sure method of smut control is by using an effective seed treatment fungicide. Spelt is extremely susceptible to loose smut and all seed should be treated with an effective fungicide.

Both seed infecting fungi contribute to poor quality seed by causing lightweight, shriveled grain. Since these seed-borne diseases are common, all small grain seeds should be treated to control seedling blight when infected seed are planted.

Soybean

Seed treatment for soybean is recommended in three disease situations; 1) when poor quality seed is used for planting (germination below 80 percent), 2) when early season control of Phytophthora damping-off is needed, and 3) when planting early into cool, wet soils, especially in reduced tillage. The major cause of poor-quality soybean seed is Phomopsis seed rot. Infected seed may be visibly moldy, yet others appear healthy. Purchasing certified seed with the percent germination, the variety name, and the seed treatment material used listed on the label is one way to ensure quality seed is planted.

Germination percentage of bin-run beans is generally unknown unless they are tested. Seed lots with greater than 80 percent germination generally do not benefit from seed treatment when planted at recommended seeding rates in warm soil. Those with less than 80 percent germination may benefit from seed treatments by increased stand and, possibly, increased yield. Decisions to use poor-quality seed should be based on the fact that any seed treatment fungicide effective against seed-borne diseases will not increase germination more than 20 percent.

If bin-run beans of 50 percent germination are treated with a fungicide, the grower should not expect over 70 percent germination in the field. Planting rates can be increased to compensate for seed of low germination. A number of fungicides are effective in controlling Phomopsis seed rot. These materials have varying degrees of activity against seed and seedling diseases.

Phytophthora damping-off is a serious disease of soybean seedlings in the more heavy. The increase in no-tillage and reduced tillage has increased the incidence of Phytophthora damping off in the state. No-till soils remain wetter longer in the spring and less precipitation is required to saturate them compared with plowed soil during the early stages of crop development. Extra moisture in the soil pore spaces favours germination of *Phytophthora*. It is recommended that seed treatments be used even when planting *Phytophthora* resistant varieties to ensure good stand establishment.

Damage from *Rhizoctonia* appears to be a stress related in that factors that adversely affect germination and establishment of seedlings predisposes the plants to infection. Stress from planting in dry soil, herbicide injury reduced tillage fields and shallow planting appears to increase the potential for damage by *Rhizoctonia*.

Adequate seed bed preparation, soil fertility, and soil moisture for rapid emergence lessens the potential for problems from *Rhizoctonia*. Sclerotinia stem rot may be introduced into new fields by sclerotia-contaminated seed lots or infested seed. Seed should be well cleaned to remove sclerotia and soybean seed should be treated with a fungicide to eradicate the fungus on infested seed.

Alfalfa and Other Small-seeded Forage Legumes

Although a number of fungicides are registered for use as seed treatments of alfalfa, clover, etc., these crops usually do not respond to seed treatments under field conditions. Some increases in stand may result but these increases have not been reflected in increased yield. The one notable exception to this is the use of fungicide seed treatment on alfalfa for control of seedling damping-off caused by *Phytophthora*. Significant improvement in stand establishment of alfalfa has been noted in research plots using treated seed.

In areas where Phytophthora root rot has caused a serious problem in reducing alfalfa stands or preventing establishment of alfalfa, a variety with resistance to *Phytophthora* that has been treated with a *Phytophthora* specific seed treatment is highly recommended.

Rhizobium Inoculation

Rhizobium and Bradyrhizobium are bacteria which can fix nitrogen from the air or collect nitrogen from the soil water solution and incorporate it into compounds that plants can use. The nodules that form on soybean roots are the sites where the bacteria reside. The bacteria may be applied to the seed or placed in the seed furrow at planting. Insecticide and fungicide seed treatments can have a negative impact on inoculants applied to soybean seed.

Several factors affect the impact of seed treatments on seed applied inoculants including: toxicity of material, toxicity of carrier or formulation, length of time inoculant is in contact with seed treatment, and formulation of inoculant. Insecticide seed treatments tend to be the most toxic of all seed treatments to inoculants. The fungicides themselves are not necessarily the problem, but the formulation or carriers may inhibit Rhizobial growth and colonisation. Using fungicides in combination with inoculants can be successful if some precautions are taken.

If a specific material is highly toxic to inoculants, then placing the inoculant directly in the furrow and not on the seed is recommended. Treating the seed first and allowing the seed treatment material to dry followed by using humus preparation of inoculant may be successful. Liquid inoculants and fungicides should not be mixed and applied simultaneously. Producers should minimise the time the inoculants are in contact with the seed treatment fungicides prior to planting.

Seed and Soilborne Diseases of Field Crops

Corn Diseases

Damping-off and seed decay. Germinating corn seed can

be attacked by several seed-borne and/or soilborne fungi. Pre-emergence and post-emergence damping-off caused by fungi is most common in poorly drained, cold soils. Seed rots and seedling blights are commonly caused by *Pythium* and *Fusarium* species, and may be caused by *Penicillium* and *Bipolaris* species. All of these fungi can rot seed prior to germination.

Figure 4. Corn seedling blight

Infected seedlings typically show a marked softening of stem tissue at the soil line. Kernels with surface cracks caused by mechanical harvesting are especially susceptible to seed rots caused by soilborne pathogens. Broad spectrum, protectant-type fungicides are highly recommended for both corn and sorghum, especially when early planting in cold, wet soils is attempted. Pythium seedling blight is most significant under these conditions. Most seed corn companies treat their seed prior to bagging. However, planter-box formulations of these materials are available for use by the grower on untreated field, sweet, and popcorn seeds.

Soybean Diseases

Phytophthora damping-off and root rot is caused by the

fungus *Phytophthora sojae*, and is particularly a problem in low, poorly drained, clay soils. However, the disease can be encountered on a variety of soil types if the soil remains wet for several days soon after planting. Phytophthora can attack soybean plants at any stage of development and stands can be reduced by seed rotting and pre-emergence damping-off. Young plants can be killed soon after emergence. Brown, water-soaked stems and yellow, wilted leaves are the primary symptoms of post-emergence damping-off. *Phytophthora* survives in the soil as thick-walled spore, called oospores.

Early in the growing season, when the soil remains wet for several days (temperatures above 60 degrees F), these spores will germinate, producing a second type of spore called a sporangium. The sporangium then germinates to produce a third type of spore called a zoospore in large numbers. Zoospores are attracted to chemicals released by soybean roots and swim through a thin film of water until a root is encountered. The zoospores produce a hyphae (a threadlike structure) and infect the soybean root. The fungus grows in the roots and eventually into the plant stem. As the plant dies, oospores are formed by the fungus.

Figure 5. Phytophthora damping off

Pythium damping-off can be caused by several species of *Pythium*. Pre-emergence damping-off, seed decays, and root rots of soybean all can be caused by this group of fungi. Pythium rots can occur on soybeans at any stage of plant development but are much more prevalent on seedlings. Diseases caused by *Pythium* can occur over a wide range of temperatures, but are most common during cold periods (below 55 degrees F) with high soil moisture. Seedlings infected with *Pythium* often fail to emerge. Stems of infected seedlings appear water-soaked and translucent above the soil line.

Diseased areas later turn brown and appear shrunken; eventually, the stem and smaller roots decay and the seedlings die. *Pythium* species survive also as oospores. During periods of high soil moisture, the resting spore germinates and infects seeds or young plants in much the same way as *Phytophthora*. Germinating soybean seeds release a wide variety nutrients or by-products that can stimulate the growth and attract spores of the fungus. Many different seed treatment materials will effectively prevent losses from Pythium seed rot and seedling blights.

Rhizoctonia seedling blight is caused by the fungus *Rhizoctonia solani,* and may occur at any time, but a protected cool period of low soil moisture followed by warming soil temperatures with a brief rainy period favours disease development. Seedlings under stress are more susceptible to Rhizoctonia seedling blight. Example of stress are herbicide injury or desiccation during early seedling growth. Seedlings and young plants are most often affected. Infected plant stems and older roots appear reddish brown and have sunken lesions.

Stems may be girdled just above the soil line; tissue thus damaged may appear cracked or cankered. In dry, windy weather, severely infected plants wilt and die rapidly. *Rhizoctonia solani* survives in the soil as sclerotia. The fungal mycelium body itself may also survive in soil or in association with old plant residue. Growth of the fungus in soil depends on soil nutrient supply, pH, moisture, and temperature. As the fungus grows in soil, it can encounter and infect germinating soybean seeds or young plants.

Figure 6. Phomopsis seed rot of soybean

Several seed treatment materials can control the seed rot phase of this disease, however, none are highly effective against the post-emergence or stem rot phase of the disease. Phomopsis seed rot is a common seed disease in and is caused by the seed-borne fungi, *Phomopsis longicolla, Diaporthe phaseolorum* var. *sojae*, and *D. phaseolorum* var. *caulivora*. Infected seeds typically germinate poorly or not at all.

Severely infected seeds appear shriveled, cracked, and elongated, and may be covered with a white, moldy growth. However, seeds may be infected and fail to show any symptoms at all. Infection of seeds is most common when warm, wet weather delays harvest. When moldy seeds are planted, seedling death results in poor stands.

Generally, higher soil temperatures favours development of seed rot and seedling blight. Seed treatments can increase the germination rate up to 20 percent. Seed lots with less than 80 percent germination should not be used.

Sclerotinia

Recent research demonstrates that Sclerotinia white mold can be introduced into new fields on infested soybean seed. In addition, this fungus, *Sclerotinia sclerotiorum*, forms hard, black, small, irregular-shaped sclerotia both inside and outside of infected soybean plants. These sclerotia are harvested with the seed or returned to the soil with plant debris at harvest. Seed should be well cleaned to remove sclerotia and treated with appropriate fungicide seed treatment to eradicate the fungus from infested seed.

Wheat and Spelt Diseases

Common bunt or stinking smut is caused by the seed-borne fungus *Tilletia laevis*. The disease causes reduced wheat yields and grain quality. Common bunt was once the most prevalent disease of wheat, however, today the disease is rarely found due to the widespread use of effective seed-treatment fungicides. Heads of diseased plants may appear stunted and are noticeably thinner and darker in colour than healthy, disease-free heads.

"Bunt balls" (kernels containing millions of fungal spores) replace the kernels in the flowering heads and spread the glumes much farther apart than normal. Bunt balls approximate the size of normal kernels but tend to be more spherical and off-coloured (typically gray-coloured). When the bunt balls are ruptured at harvest, they release spores into the surrounding air and contaminate the surface of healthy seed.

Bunt fungi survive as resting spores on contaminated seed. When contaminated seeds are planted, bunt spores germinate in the presence of moisture and infect the wheat seedlings. Once established, the fungus grows systemically in the plant. At flowering, entire flowering heads may be infected, and the fungus converts the developing kernels into bunt balls. Healthy seeds are contaminated by the release of the spores at harvest.

Figure 7. Common bunt of wheat

Loose smut, caused by the fungus *Ustilago tritici*, has little effect on seed quality. However, it can result in substantial yield losses if left unmanaged. Symptoms are most noticeable between heading and maturity of the crop. Diseased flowering heads are conspicuously darker when compared to healthy, green heads. Also, diseased heads typically emerge earlier. Diseased spikelets may be entirely transformed into dark fungal spore masses. As the head emerges, floral tissue is torn and the spores are released.

Eventually only a spike is left where a normally healthy head should be. Spores of the fungus are dispersed by wind to nearby healthy flowers where they initiate new infections during wet weather. Unlike bunt, the loose smut fungus infects the developing kernels without causing noticeable damage.

Figure 8. Loose smut of wheat

Ustilago tritici survives in infected wheat seed. The fungus becomes active when the seed germinates and grows into the growing point of the developing wheat plant. Fungal growth closely follows the plants' growth. When the flowers form, the fungus again sporulates and starts the cycle again. Use a systemic fungicide seed treatment to eliminate this pathogen from infected seed.

Infected seed appear whitish to reddish in colour and are nearly always shrunken and light in weight. Scabby kernels may be dead or the young seedling may be killed by seedling blight after emerging from the soil. Roots of plants killed by seedling blight appear light to reddish brown in colour and may be covered with mold. If they survive, they generally lack vigour and may produce only a few weak tillers. Planting infected seed early when soil temperatures are above 60 degrees F generally increases loss from seedling blight.

The best control of the seed-borne phase of scab is to first clean seed lots to remove all lightweight seed, thus increasing test weights; and second, treat with a fungicide that is effective in controlling seedling blights.

Figure 9. Head scab of wheat

Stagonospora glume blotch, caused by the fungus *Stagonospora nodorum* has increased in the US in recent years probably due to expanded use of nitrogen fertilisers, semidwarf wheat varieties, and reduced tillage. Symptoms can develop throughout the growing season on all above-ground plant parts. Initial symptoms are small chlorotic flecks usually on lower leaves or those in contact with soil. The glumes of the wheat heads become infected in late spring before flowering. Lesions generally begin at the glume tips and are dark brown in colour.

During favourable weather, the fungus penetrates the glumes and infects the seed causing severe shrivelling of the grain. Stagonospora survives on seed, straw, and/or volunteer wheat. With the onset of moist weather, spores are produced and are spread to healthy wheat plants by splashing rain. Glume blotch is most common in relatively warm weather when the temperature is between 70 degrees F and 80 degrees F.

Barley Diseases

True loose smut of barley is caused by the fungus *Ustilago tritici*. The disease cycle is exactly the same as that of loose smut of wheat. Infection occurs when the flowers develop. The seeds become infected internally and must be treated with a systemic fungicide to eliminate the pathogen. Semiloose smut is caused by the fungus, *Ustilago avenae*. The disease typically causes small losses due to the use of resistant cultivars and fungicide seed treatment. It is most common when soil temperatures are cool and the soil is dry.

Distinguishing semi-loose smut from true loose smut in the field is extremely difficult. Diseased heads typically appear earlier than healthy heads. Millions of fungal spores may be formed in diseased heads. *Ustilago avenae* survives as teliospores on the seed surface. At seed germination, the fungus becomes active and infects young seedlings prior to their emergence from soil. As the plants develop, the fungus grows within the growing point of the young plant. At the boot stage, the fungus transforms the kernels and glumes into dark spore masses very similar to those produced by the true loose smut fungus.

Spores are readily dispersed by wind or the action of harvesting equipment. Healthy seeds are contaminated upon contact with the spores. Covered smut caused by the fungus, *Ustilago hordei*, can result in significant yield losses if left untreated. It is a relatively common disease throughout the barley-growing regions of the world. Problems with the disease occur over a wide range of temperatures and are most common under moist soil conditions.

Ustilago hordei survives as thick-walled resting spores on barley seed. Typically, spore germination coincides with seed germination, it is at this time that infection occurs. Once established, the fungus invades the actively growing tissue and keeps pace with growth of the plant. At flowering, the fungus grows through the floral tissue and forms masses of spores in place of healthy seed. Each kernel becomes a spore filled "smut ball" that is enclosed by a thin membrane until plant maturity. At harvest, spores are liberated and may contaminate healthy seed.

Barley stripe, caused by the seed-borne fungus, *Drechslera graminea*, is most common during periods of rain or high humidity. Few seeds are produced on infected plants, and losses generally coincide with disease levels in a particular field. Properly treated seed is the primary means of controlling this disease. Symptoms are first noticeable on the first two or three leaves produced by the plant and on many leaves produced later. A yellow stripe on the leaf sheath and basal portion of the leaf blade is common.

Later, the stripe may extend the entire length of the leaf and the leaf may die. At heading, diseased plants are often stunted and appear a light tan colour in contrast to the larger, green healthy plants. Spikes may be deformed or may fail to emerge. At heading, spores are produced on diseased leaves and spread by wind to nearby heads. Grain produced by diseased plants is typically brown and shriveled.

Drechslera graminea survives entirely as mycelium in the outer layers of the seed. When the seed germinates in moderately moist soil at temperatures below 55 degrees F, the fungus mycelium penetrates and infects the seedling. The fungus grows within plant tissue and occupies the actively growing plant parts.

Net blotch and spot blotch, common diseases of barley, are caused by the fungi, *Drechslera teres* and *Bipolaris sorokineana,* respectively. Net blotch is most common during rainy and humid weather, and early flag leaf infections can result in both reduced grain yield and weight. A net-like pattern on sheath leaves or flag leaves is the most recognisable symptom of the disease. Initially, small spots or streaks appear on the leaves, followed by an expansion into conspicuous longitudinal and transverse streaks, eventually combining into a network. Brown areas surrounded by greenish-yellow margins are commonly observed on infected leaves. Eventually, entire leaves may wither and die.

Drechslera teres survives in infected seed or on crop residues. Infection of seedlings is most common at 34 degrees F to 60 degrees F. Once infection is established, the fungus produces spores on leaf tissue during periods of high relative humidity. Spores are released and carried by wind to nearby barley plants where new infections can result. Infection of leaves can occur when leaves are wet for periods of 5 to 15 hours within a 46 degrees F to 91 degrees F temperature range.

Spot blotch is most commonly a problem during warm, humid weather. Yield losses of up to 36 percent have been attributed to this disease when the flag leaf has become severely diseased early in the season and has died prematurely. Seedling blight (*B. sorokineana*) in winter barley is favoured by planting early in warm soil. Delayed planting of winter barley in cooler soil helps prevent blight.

Small brown spots surrounded by a yellow halo is the characteristic symptom of the disease. Both leaves and leaf sheaths on plants of all ages can become infected. Eventually, the spots coalesce and cover large areas of the leaves. Minute fungal fruiting structures containing spores can sometimes be observed in the larger spots. Severely diseased leaves may eventually die.

Bipolaris sorokineana survives in seed or infested crop residue. Leaf infection results from airborne spores produced on the seed or on residue. Epidemics of spot blotch can occur when there are prolonged wetness periods (greater than 16 hours) and the temperature is above 70 degrees F. Note that net blotch is more common in cooler, humid weather, as opposed to spot blotch which prefers warm, humid weather. Most common broad-spectrum seed treatment fungicides are effective in eliminating seed-borne *B. sorokineana* and *D. teres* as well as preventing losses from seedling blight caused by them.

Oat Diseases

Loose smut and covered smut, caused by *Ustilago avenae* and *U. segetum*, respectively. When disease outbreaks are severe, both yield and grain quality can be significantly reduced. However, fungicide seed treatments effectively control both diseases. Symptoms of these diseases are very similar. Smutted oat panicles emerge from sheaths as olive-brown to black fungal spore masses.

Figure 10. Loose and covered smuts of oats

Smutted panicles remain more compact than healthy ones, and usually all spikelets on a diseased plant are affected. Smutted plants are often stunted and easily detected in a recently headed field of oats. The thin membrane enclosing the spore masses of *U. avenae* usually rupture and disintegrate after panicle emergence as opposed to those of *U. segetum*, which can persist longer.

Alfalfa Diseases

Phytophthora damping-off and root rot is caused by the fungus, *Phytophthora megasperma f.sp. medicaginis*. Standing water, poor drainage, and generally wet soil conditions favour disease development. This disease can be lethal to seedlings and established plants. Seedlings may fail to emerge or can be killed after emergence. A conspicuous yellowing of leaves, particularly lower ones, is a characteristic symptom. Infected seedlings often wilt. Seedling tap roots have dark brown to black lesions and may be rotted 2-4 inches below the crown.

A yellow discoloration of internal root tissue is often present. Tap roots may appear discoloured. The life cycle of *Phytophthora* on alfalfa is similar to that on soybean, however, the fungi are completely different and do not attack the other plant host. Seed treatments are labelled for control of Phytophthora damping off. However, to protect stands beyond the seedling stage, alfalfa varieties with resistance to the fungus should be used.

10

Blind Seed Diseases

During the early 1920s, growers of perennial ryegrass (*Lolium perenne* L.) seed in New Zealand were troubled by poor germination of their seed crops. A systematic inquiry initiated in 1923 associated reduced germination with humid conditions during seed development, but the cause was unknown. By 1926, germination was as low as 19 percent, and in the southern region of New Zealand 95 percent of the seed lots tested had germination of 90 percent or less. Ungerminable seed had an abnormal appearance characterised by opacity, roughness, and a reddish caryopsis surface. These symptoms were difficult to see unless the lemma and palea, which cover the caryopsis, were removed. The difficulty in visual detection of the ungerminable (diseased) seeds led Neill and Hyde to propose "blind seed" as the common name of the disease.

By 1932, it was apparent that a conidia-producing fungus was associated with affected seeds, but it was not until 1937 that pathogenicity of the blind seed fungus (tentatively identified as a *Pullularia* sp.) was established. However, another fungus, distinct from *Pullularia,* was also found associated with infected seed. After careful study, this second fungus, not *Pullularia,* was found to cause blind seed disease.

In 1942, after an investigation of the life history of the blind seed fungus, Neill and Hyde determined that a fungus called *Phialea temulenta* Prill. & Delacr. was identical to the blind seed fungus on ryegrass in New Zealand. *P. temulenta* was previously reported on seed of rye (*Secale cereale* L.) in France in 1891. In 1945, Wilson et al. reviewed the taxonomic placement of *P. temulenta* and erected a new genus, *Gloeotinia*, to accommodate it.

The effects of blind seed disease on the production of grass seed can be tremendous. Germination in infected seed samples has been as low as 1 percent in New Zealand, 13 percent in the United States, and 50 percent in Great Britain. Blind seed continues to periodically plague growers in New Zealand, and its recent reappearance in the United States has renewed interest in the disease here.

The blind seed fungus was first recorded on infected seeds of rye (Secale cereale L.) in France in 1891. Although first reported on rye, its subsequent occurrence on this crop is very rare. Blind seed disease is primarily a problem of forage and turf grasses grown for seed.

Blind seed disease was unknown in Great Britain until after its discovery in New Zealand. However, the connection between blind seed and low germination in ryegrass (Lolium sp.) was suspected to be of long standing in Great Britain, since low germination in some years was well known. Proof of the long-standing occurrence of blind seed was established when conidia of G. temulenta were found among stored seeds from a 1909 ryegrass crop grown in Ireland.

The identification of blind seed disease in the United States in 1944 established that the fungus was widely distributed on ryegrass grown for seed, a distribution likely established through the international grass seed trade.

Blind seed has been reported from Australia, including Tasmania, Victoria, and New South Wales; Denmark; England, including Kent, Sussex, Hereford, and the Isle of Man; Ireland; France; The Netherlands; New Zealand; Northern Ireland; Scotland, including Ayrshire and the Shetland Islands; Sweden; United States, including Oregon; and Wales.

Worldwide, 56 host species have been reported as susceptible to G. temulenta. Most hosts are in the subfamily Pooideae, tribes Avenae and Poeae, with heaviest infections reported in the genera Agrostis, Festuca, Lolium, and Poa. Lolium perenne is widely recognised as susceptible and has been identified as a host from all countries reporting blind seed disease. In the Triticeae, moderate to heavy infections were observed on Psathyrostachys, Pseudoroegneria, and Secale species.

Grasses in the Bromeae appear less susceptible, with little to no infection observed among species of Bromus. In the United States, blind seed disease was found on species of Agrostis, Aira, Alopecurus, Bromus, Cynosurus, Deschampsia, Danthonia, Festuca, Glyceria, Hordeum, Holcus, Lolium, Phleum, and Poa. Despite the susceptibility of many common grasses in the United States, G. temulenta has been reported only from Oregon. In New Zealand, blind seed was reported on Agrostis, Cynosurus, Festuca, Holcus, Lolium, Poa, and Secale cereale.

In Northern Ireland, blind seed was found on Agrostis, Cynosurus, Festuca, Holcus, Lolium, and Poa. Additional host reports include Calamagrostis from Germany, Elytrigia from Norway, and Secale from France and Germany. Most of these U.S. hosts were reported in a comprehensive host range study by Hardison. However, there is one discrepancy in the U.S. host range. Fischer reported G. temulenta on Danthonia Californica Boland (subfamily Arundinoideae, tribe Danthoneae). Hardison, however, did not observe infection on D. Californica inoculated with G. temulenta under natural or artificial conditions. Additional studies are needed to determine all grasses that are susceptible to G. temulenta and their relative susceptibility.

In the production of grass seed, loss from blind seed disease occurs through a reduction in germinable seed since infected seeds are ungerminable. In addition, seed lots with germination below certification limits, or below seed contract standards, are of less value and in some countries are unmarketable. Presence of G. temulenta in import seed shipments may result in rejection of the seed by some countries.

Symptoms

The seed is the only component of the host plant infected by G. temulenta. Infected caryopses appear shriveled, rough on the surface, and rusty brown or pinkish in colour. Conidia accumulate on the seed surface in a spore secretion (slime), which may be waxy and clear or pale pink in colour or may appear as a reddish-brown crust. Healthy caryopses normally appear golden brown, plump, and smooth. However, infected seeds covered by the lemma and palea are difficult to discern from normal seeds. A consequence of blind seed infection is reduced germination, and the correlation between percentage of infected seed and percentage germination in ryegrass is well established. Germination of infected seed is rarely greater than 10 percent.

Technical Description

Stroma: Infection of the grass caryopsis results in the mummification of the caryopsis, creating a substratal stroma. Hyphae, 3-4 µm wide, ramify throughout the pericarp, teste, and endosperm and are not differentiated into rind and medullary parenchyma. A true sclerotium does not develop, although the infected seed functions similarly to a sclerotium as a means of survival through the winter.

Sporodochia. In late winter or early spring, pinkish, pulvinate, gelatinous sporodochia form either on the surface of the pales or between the pales and caryopsis. They are 0.4-1 × 0.5-1.5 mm in size. Sporodochia consist of a core of closely septate, branching hyphae with the terminal cells of each branch bearing 1-4 microconidiophores.

Microconidiophores and microconidia (spermatia). Microconidiophores are 2-5 µm in diameter and 5-9 µm long, septate, guttulate, hyaline, and penicillate (branched 2 or 3 times). Microconidia are first formed by a constriction below the apex of the microconidiophore. The rest bud off in succession inside a tube formed by the terminal portion of the microconidiophore. Microconidia are unicellular, uninucleate, ovoid, guttulate or biguttulate, hyaline, 1.8-3.0 × 2.3-6.0 µm. In microconidial germination, a terminal germ tube forms; or if a transverse septum forms, a terminal or lateral germ tube will be produced.

Macroconidiophores and macroconidia. Macroconidiophores are short barrel-shaped cells, 2-3 µm wide and 5-15 µm long, that arise laterally on the hyphae. Macroconidia are budded from the apex of the macroconidiophores and arrange in clusters perpendicular to the hypha. Up to 30 macroconidia develop per conidiophore. Macroconidia are smooth, unicellular, uninucleate, hyaline, cylindrical to slightly cresentric with rounded ends, and usually biguttulate. They are 2.5-6 × 11-21 μm in size. The vegetative nucleus is 3-5 × 2 μm and the nucleolus may be as large as 2 μm. On the surface of the caryopsis, macroconidia are embedded in a pinkish, slimy mass that dries to form a hard reddish-brown crust. When germinating, macroconidia swell and produce one or two germ tubes.

Apothecia. Apothecia are small, fleshy, and cup-shaped. One to 7 (usually 1 to 3) apothecia emerge from each infected seed. The stipe is smooth, velutinous under magnification, externally white or gray, internally pinkish brown, enlarging upward, and longitudinally furrowed. The stipe varies from 1 to 10 μm in length and from 0.2 to 0.5 μm in diameter and is composed of hyaline, parallel hyphae, 4-6 μm in diameter, occasionally intertwining and seldom branched. Apothecia emerge from the caryopsis and elongate.

The disk of the apothecium is at first closed but opens to cup-shaped and with age becomes saucer-shaped and then flat. The disc diameter ranges from 1.0 to 7.0 μm. The disk colour changes from light pinkish brown to deep brown, orange brown, or pale pinkish cinnamon, darkening to cinnamon when old. The margin is smooth and entire and is radially wrinkled around the stipe apex.

Hymenium. The hymenium is 100-140 μm deep. The subhymenium consists of intricately intertwined and coiled hyphae 2.5-3 μm in diameter. The subhymenium blends into the medullary excipulum, a 22-27 μm deep layer composed of fine, densely intertwining hyphae 2-5 μm broad. The outermost layer (the ectal excipulum) is 35-40 μm thick and is composed of parallel to somewhat interwoven hyphae 3.5-4.5 μm in diameter.

Asci. The asci are cylindrical and clavate, with 8 spores obliquely placed in a single row (uniseriate) in the upper two-thirds of the ascus. Ascus size is variable but falls within the range of 66-120 µm long × 3-8 µm wide. The ascus base tapers to about 2-5 µm. The apical cap is 1-3 µm thick, and the apical plug does not stain blue with iodine.

Ascospores. Ascospores are hyaline, smooth, elliptical, fusoid to broadly fusoid, and usually biguttulate. One side is often flattened, or curved, continuous, or rarely developing a central septum. Ascospore size is variable, 7-14 × 2.5-4.5 µm. Germinating ascospores swell to about 10 × 5 µm. The first germ tube is terminal, followed by a second that is frequently lateral in position and usually constricted at the point of origin. They normally develop a central septum and two polar hyphae, but often lack a septum and have a single polar or lateral hypha.

Paraphyses. Paraphyses are fusiform, hyaline, nonseptate or sparsely septate and 1.5-4 µm wide. Spooner described the paraphyses as enlarging at the apex to 2.5-3.0 µm, but others reported that the apex was not swollen. Paraphyses are as long as or slightly longer than the asci.

Growth

On a nutrient medium such as potato dextrose agar, G. temulenta grows slowly and produces a partly submerged, branching, hyaline, septate mycelium. Sporulation and slime production occur after 7 days and in culture appears reddish brown or chocolate brown. The addition of 1-percent peptone to PDA or malt agar increases spore mucilage production. However, some cultures are predominantly mycelial while others are conidial. In culture, macroconidia are produced from short conidiophores formed at intervals perpendicular to the hypha.

Conidia from culture may be larger or appear less regular than those from seed. Growth is slow at 5°C, optimal at about 20°C, less at 27°C, and restricted at 30°C. Radial growth slows with decreasing water potential through -9.0 to -1.0 MPa. Sporodochia develop in culture at 5°C to room temperature after about 1-3 months.

Biology and Epidemiology

Apothecia

The overwintering, or survival, unit of G. temulenta is the infected seed. Infected seeds reach the soil by shattering, by seed loss during harvest operations, by planting of diseased seeds, and by natural seed dispersal in harvested areas. Infected, ungerminable seeds resist attack by bacteria and molds and do not decay as they overwinter. At or near the soil surface, G. temulenta continues to develop within the seed. Moist soil conditions with temperatures near 2°C for about 8 weeks are required to induce the sexual (apothecial) stage of G. temulenta.

The precise biochemical changes that occur or metabolic pathways affected during this conditioning have not been determined. In spring or early summer, at or prior to flowering of perennial ryegrass, apothecia emerge from the overwintering infected seeds. Usually one to three, but as many as seven, apothecia can emerge from a single infected seed. Not all infected seeds will yield apothecia. In fact, only 5-30 percent of ungerminated seed produce apothecia.

Ascospores

Large numbers of ascospores are ejected from each

apothecium in response to slight changes in relative humidity. Ascospores that land on flowers, including the stigma, ovary, or styles, will germinate and infect the host. However, seeds can be infected up to the time they reach their maximum size.

Within about 7 days to 16-17 days after inoculation, the conidial stage is manifest—a pinkish slime in which conidia are embedded. These spores are relatively short-lived, about 1 month. However, a few conidia may survive as long as 4-6 months if stored under cool, dry conditions.

Wet seasons, especially during anthesis in the grasses, are clearly supportive of blind seed infection. Large numbers of apothecia can appear during wet weather. Wet seasons, combined with low temperatures, extend the period of apothecial production and spore release. However, not all apothecia are produced at the same time. Some apothecia develop early, others late. Under cool (13°C), wet conditions, apothecia can be produced over a 2-month time frame. The expected lifespan of an individual apothecium is about 8-14 days, although they shrivel within a few hours in a dry atmosphere.

Temperatures of 10-16°C and high humidity are considered ideal for blind seed development. Infection does not occur under very warm (30°C) temperatures. Blind seed disease could spread from infested areas to noninfested areas, based on observations of commercial fields and field plots planted with pathogen-free seed. Additional sources of infection include seed for pastures, hedgerows with susceptible grasses, and waste ground.

Direct observations of spore movement were made by Neill and Armstrong, who trapped spores of G. temulenta 18 m high and at ground level 1.6 km from the nearest infected field. The highest rate of infection occurs while florets are open. The potential for infection reduces greatly after flowering. Cool, moist weather conditions aid dispersal, prolong the period of pollination, and extend the period of greatest susceptibility of the plant.

Flowering in a ryegrass spike begins at the top and progresses downward over about 10 days. Production of conidia begins within 6 days of infection and increases for about 16 days. Consequently, infection of upper florets by windborne ascospores may result in the spread of subsequently produced conidia to lower florets under rainy conditions. Rain dissolves the slime in which conidia are embedded and provides a vehicle for their secondary spread.

Involved with transmission of the condidial slime. However, no observations or data on association of Gloeotinia with insects or their ability to vector G. temulenta has been published. Infections occurring at flowering or prior to endosperm formation resulted in seeds that are thin and light in weight. These infected seeds may not be capable of supporting apothecial production, although they may support development of macroconidia.

Abundant production of macroconidia during early flowering or seed development provides inoculum for secondary spread and subsequent disease development. Seeds infected during the early to middle stages of development are approximately normal size and weight, and a large quantity of spores are produced. Seeds infected late in development may be capable of germination. Fewer spores are produced from late infections than from early ones. The potential for rapid increase in blind seed severity was emphasised by Hardison, who noticed a rapid increase in disease over a 1- to 3-year period.

Fungal Genetics and Physiology

G. temulenta is heterothallic—it requires genetic exchange between two different mating types for sexual reproduction and subsequent production of apothecia. G. temulenta has two mating types that are identical in all morphological features. Within each apothecium half of the ascospores are of each mating type, arbitrarily called "a" and "b." Apothecia will develop only after mating types a and b come into contact with one another and undergo fusion.

Conidia produced following infection from an ascopore of one mating type will produce only conidia of that mating type. Genetic exchange between types can occur through transfer of macroconidia from one infected seed to another or through transfer of microconidia, which can develop on the seed in spring after the seed has overwintered. A conjugation tube—a device to exchange genetic information—can form between pairs of macroconidia even before either conidium germinates. As expected from the heterothallic requirement of G. temulenta, relatively few infected seeds produce apothecia.

The vegetative hyphae are uninucleate. Chromosome number in G. temulenta is n=15, and mitotic chromosomes range in size from 0.25 to 1.0 μm.. In the microconidiophores the nucleolus is lacking, RNA is low, and the level of RNA depends on the level in the subtending cells. Microconidia have not been observed to germinate and produce a vegetative mycelium but can serve a sexual function.

Little is known about variability in virulence of G. temulenta. Sproule and Faulkner reported variation in aggressiveness among strains of G. temulenta. Wright and Sproule reported that disease ranking of clones was the same when mixed blind seed isolates from The Netherlands or the British Isles were used. Little is known about the physiology of G. temulenta. A cold conditioning period of about 8 weeks is required to induce the apothecial phase. The metabolic pathways or mechanism associated with the induction have not been investigated.

Disease Control

The survival propagule of G. temulenta is the infected seed. Control measures centre around removing as many infected seeds as possible from the field during harvest and avoiding introduction of infected seed by using disease-free or treated seed. Maintaining a healthy stand through good fertilisation practices also contributes to control of blind seed. An integrated approach to blind seed control should consider disease resistance, field location, seed source, seed treatments, planting, time of closing, fertilisation, stand density, fungicide sprays, methods of harvest, postharvest residue management (straw residue removal, postharvest plowing, crop rotation, field burning), and postharvest seed cleaning.

Disease Resistance

The search for resistance to blind seed began shortly after discovery of the disease. Early investigations in New Zealand compared indigenous grasses to commercial grasses. Differences in susceptibility were attributed to timing of flowering and favourability of climatic conditions during flowering. Early attempts at breeding ryegrass for resistance to G. temulenta were confounded by high variability and inconsistent results. Resistance was quantitative and repeatable across environmental conditions and fungal strains. Unfortunately, little resistance is believed to be present in most cultivars of perennial ryegrass and tall fescue now grown commercially for seed.

Locating fields away from infested fields to avoid the introduction of inoculum from nearby sources is recommended. To prevent establishment and persistence of infected seed, grazed areas not kept for seed should be topped when seed heads appear. Surrounding fields with crops such as cereals or root or forage crops may provide a barrier to movement of spores into a field, although long-distance (more than 1 km) airborne movement of ascospores can occur. Since infected seed is the source of inoculum, planting disease-free seed is recommended.

Treatments

G. temulenta has limited survival in seed stored dry. Some researchers controlled blind seed with a hot water treatment that included either a 4-hour pretreatment with tepid water, then 15 minutes at 50°C, or no preimmersion treatment and 30 minutes at 50°C. The treatments provided full control with little or no reduction in seed germination. After hot water treatment, infected seeds decayed in the soil. Untreated infected seeds resisted decay.

Numerous fungicides have been evaluated for their efficacy as seed treatments for blind seed disease. Although Hair reported some success, most of the early research indicated that chemicals applied as seed protectants were not effective against blind seed disease. However, modern systemic fungicides such as benomyl have proven effective as a seed treatment. In New Zealand, seed treatment with fungicides has proven effective and is recommended for control of blind seed disease.

Planting

When seeds are planted more than one-half inch deep, apothecia have difficulty reaching the soil surface. Good preparation of the seed bed facilitates planting at the proper depth and good coverage of seed. Fields with heavy soils or poor drainage may be more favourable for blind seed development because they provide the prolonged moist conditions that are favourable for production of ascospores. Good soil drainage provides conditions that are less favourable for apothecial production.

Infected seed must undergo a cool, moist period for about 8 weeks to induce the reproductive (apothecial) phase of the pathogen. Wright found that when seed was planted in spring, apothecial production did not occur; the requirement for cold conditioning was not met. Planting a susceptible first-year companion crop such as L. temulentum is not recommended because of its potential to increase inoculum if seed becomes infected.

Nitrogen Fertilisation

Numerous studies indicate a reduction in blind seed in response to manure or nitrogen fertilisation. Chestnutt and Rutherford reported a significant reduction in blind seed in manured plots, compared with unmanured plots of perennial ryegrass. Lynch and Lithgow and Cottier observed that nitrogen improved yield and germination, although the effect of nitrogen on blind seed was uncertain. In a paired-plot experiment, Stewart found blind seed levels decreased in plots treated with nitrogen compared with untreated plots.

Hampton and Scott established that a decline in blind seed between 1960 and 1980 in New Zealand correlated with the increased use of nitrogen fertiliser. In field trials, they demonstrated that as nitrogen rate increased, the rate of blind seed infection decreased, a result also reported by Hampton and de Filippi et al.. Under laboratory conditions, Hampton and Scott observed that urea directly suppressed apothecial formation. However, in field plots, Hampton and Scott found no significant differences in number of apothecia among field plots treated with various levels of urea, although a reduction in blind seed infection was observed in urea treatments. They concluded that nitrogen fertilisation altered the physiology of the plant, enhancing resistance to G. temulenta.

In subsequent studies some other researchers examined the level of blind seed in adjacent irrigated and nonirrigated field plots to which various rates of urea had been applied. In irrigated field plots, nitrogen application significantly reduced blind seed disease, but this did not occur in nonirrigated plots. As the inoculum source was external to the trial, they concluded that plants which are able to utilise available nitrogen develop a greater capacity to resist blind seed. The mechanisms associated with this resistance need to be determined.

In addition to increasing resistance, nitrogen applications can also increase lodging or increase stand density, providing a physical barrier to restrict spore movement up through the canopy.

Fungicide Sprays

Fungicides such as tebuconazole or carbendizim are effective for blind seed control. Sprays applied as soil drenches or to the soil surface have been shown effective in reducing the number of apothecia. Benomyl applied at 2.8 and 5.6 kg/ha reduced apothecia 80 and 90 percent, respectively.

Harvest

Since the primary source of inoculum is the infected seed, early harvest to avoid excessive seed shatter is recommended. Osborn suggested early harvest under dry conditions as a source of disease-free seed, since late season disease could develop with a change in the weather to wet conditions. Removal of lightweight or infected seeds during harvest reduces inoculum left in the field.

Straw Residue Removal

Since dry soil conditions are unfavourable for apothecial development and spore release, some researchers recommended removing the straw after harvest to allow the soil surface to dry more rapidly in spring. In Oregon, residue is commonly baled and removed from the field. In some cases the straw is finely chopped with specialised flails. Residues that are not sufficiently chopped decompose slowly and can interfere with crop growth or development and may leave the soil wet for prolonged periods.

Postharvest Plowing

Plowing infested fields reduces the area of infestation by burying much of the inoculum source—the infected seeds.

Field Burning

Excellent control of blind seed is achieved with postharvest field burning. For optimal control, the entire dry-straw residue should be open burned. Burning by propane flaming after residue removal (baling) is not as effective as open burning, since propane does not achieve the temperatures of open-grass burning.

Seed Cleaning

Recleaning of seed lots is not very effective in reducing the level of blind seed. Cleaning to a higher seed weight by removing infected seeds improved germination for some seed lots with a high level of infection; but in lots with a low level of blind seed, cleaning simply removed small but viable seed. A relationship between seed weight and germination could not be established. Since infected seed are present in screenings, destroying the screenings destroys the inoculum.

Bibliography

Bhale, M.S., D. Khare, N.D. Raut and D. Singh, *Seed-borne Diseases Objectionable in Seed Production and Their Management*, Jodhpur, Scientific, 2001.

Brunt, J.A., ed., *Seed Pathology and Microbiology*, Vols. 1-11, Annual Bibliography, CAB International, Wallingford, Oxon, UK and DGISP, Denmark, 1990-2000.

Chaube H.S., and Ramji Singh, *Introductory Plant Pathology*, Lucknow, IBDCO, 2001.

Dixit Festschrift S.N., edited by K.G. Mukerji, et. al., *Current Concepts in Seed Biology*, 1992

Gautam N.C., *Plant Biotechnology*, New Delhi, Shree Pub., 2006.

Jayabalan N., *Plant Biotechnology*, New Delhi, A.P.H., 2006.

Khan, Aslam, *Plant Anatomy and Physiology*, Delhi, Kalpaz, 2001.

Lela V. Barton, *Seeds : Their Preservation and Longevity*, Delhi, Asiatic Pub., 2004.

Lokeswara Rao M., *Plant Biodiversity of World*, Dehradun, International, 2004.

McDonald M.B., and L.O. Copeland, *Seed Science and Technology : Laboratory Manual*, Reprint. 1999.

Reddy, S.M., *Frontiers in Microbial Biotechnology and Plant Pathology*, edited C. Manoharachary, D.K. Purohit, S. Ram Reddy, M.A. Singara Charya and S. Girisham. Jodhpur, Scientific, 2002.

Sharma, Rajni, *An Introduction to Plant Morphology*, New Delhi, Campus, 2004.

______________, *An Introduction to Plant Physiology*, New Delhi, Campus Books, 2004.

______________, *An Introduction to Plant Anatomy*, New Delhi, Campus, 2004.

Siddiqi, S., ed., *Seed Pathology and Microbiology*, Vols. 12-13, Annual Bibliography, CAB International, Wallingford, Oxon, UK and DGISP, Denmark, 2001-2002.

Sudhir M., *Plant Biotechnology*, New Delhi, Dominant, 2004.

Tribhuwan Singh and Kailash Agrawal, *Seed Technology and Seed Pathology*, Jaipur, Pointer Publishers, 2001.

Index